Chemical Separations in Nuclear Waste Management

The State of the Art and a Look to the Future

EDITORS

Gregory R. Choppin

Department of Chemistry and
Institute for International Cooperative Environmental Research
Florida State University
Tallahassee
Florida, U.S.A.

Mikhail K. Khankhasayev

Joint Institute for Nuclear Research
Dubna, Russia, and
Institute for International Cooperative Environmental Research
Florida State University
Tallahassee
Florida, U.S.A.

Hans S. Plendl

Department of Physics and
Institute for International Cooperative Environmental Research
Florida State University
Tallahassee
Florida, U.S.A.

ASSISTANT EDITOR
Elizabeth Purdum

DOE/EM-0591

CHEMICAL SEPARATIONS IN NUCLEAR WASTE MANAGEMENT

The State of the Art and a Look to the Future

Editors

Gregory R. Choppin
Mikhail K. Khankhasayev
Hans S. Plendl

Institute for International Cooperative Environmental Research
Florida State University

 Battelle Press

Columbus • Richland

This book is a publication (DOE/EM-0591) of the U.S. Department of Energy. The publication of the book was supported through Cooperative Agreement DE-FC26-01NT40842 between the Department of Energy and the Florida State University.

Library of Congress Cataloging-in-Publications Data

Chemical separations in nuclear waste management : the state of the art and a look to the future / editors, Gregory R. Choppin, Mikhail K. Khankhasayev, Hans S. Plendl ; assistant editor, Elizabeth Purdum.
  p. cm.
  Papers presented at an international workshop held in Prague on Sept. 11, 2000.
  Includes bibliographical references and index.
  ISBN 1-57477-121-3 (alk. paper)
    1. Radioactive wastes—Management—Congresses. 2. Separation (Technology)—Congresses.
  I. Choppin, Gregory R.   II. Khankhasayev, Mikhail K   III. Plendl, Hans S.

TD898.14.M35 C493 2002
621.48'38–dc21   2001043590

Printed in the United States of America

Battelle Press
505 King Avenue
Columbus, Ohio 43201-2693
614-424-6393; 1-800-451-3543
Fax: 614-424-3819
Web bookstore: www.battelle.org/bookstore
E-mail: press@battelle.org

CONTENTS

PREFACE

Chemical separations in nuclear radiochemistry science and technology are concerned with the separation of mixtures of materials according to certain properties. The separation of nuclear materials into their chemical components is at the heart of all efforts to reduce significantly the volume of high-level nuclear waste inherited from the production of nuclear weapons during the Cold War years and from continued production of nuclear energy. This waste is associated with serious long-term health risks, enormous environmental contamination, and potential contamination of unpolluted areas. The reduction of the nuclear waste to manageable levels is essential for nuclear energy to remain an option for meeting the energy needs of the future. The separation of radionuclide species in nuclear waste from each other also makes it possible to treat long-lived radionuclides differently from shorter-lived ones, resulting in more options for present and future disposal.

This volume evolved from an international workshop, Technologies for Nuclear Separations: A Look to the Future, held in Prague on 11 September 2000 prior to the Fifth International Symposium and Exhibition on Environmental Contamination in Central and Eastern Europe. During this workshop, scientists from several countries discussed the history, present state, and possible future directions of separation science and technologies that may lead to a cleaner, safer, and healthier 21st century. This volume is based on the papers presented at that workshop and on the discussions following the presentations. The authors updated and augmented their original papers for this publication. For readers who are not experts in separation science, the evolution of this science from the onset of nuclear research at the beginning of the last century to the present is presented to help them understand its foundation, contributions, and significance. Readers are also provided an overview of the enormous environmental legacy produced by the nuclear weapons complexes in the United States and in the former USSR. They are made aware of the magnitude of the cleanup efforts that are underway in both countries and of the pivotal role played in these efforts by separation science.

The chemistry of actinides is fundamental to understanding chemical separation methods of nuclear materials containing any of these elements. This group, from actinium to lawrencium, contains the reactor fuels thorium, uranium, and plutonium as well as the "minor" actinides—neptunium, americium, and curium—that are produced by neutron capture in reactor operation. Actinide chemistry is, therefore, at the heart of many techniques of separation science. Since actinides differ in several important respects from other elemental series and are not generally discussed in detail in university chemistry courses, their chemistry is summarized in the introduction to this volume.

The editors thank Dr. Roy Herndon, Director of the Institute for International Cooperative Environmental Research, Florida State University, for actively supporting the publication of this volume and for initiating the international workshop from which it originated. They also thank Professor Jan John of the Czech Technical University in Prague for his many contributions to the success of the workshop.

EXECUTIVE SUMMARY

On 11 September 2000, an international workshop on Technologies for Nuclear Separations: A Look to the Future was held in Prague, Czech Republic. This workshop was convened to review and discuss the state of nuclear separation technology development in the context of future needs and opportunities. At the workshop some of the world's leading nuclear scientists discussed how separation technologies may develop in the future and how they may lead the way to a cleaner, safer, and healthier 21st century. This book grew out of that workshop, and it is intended to inform graduate students in basic and applied sciences, professionals in the fields of nuclear waste management and nuclear power generation, and other interested individuals about the history, the present state, and the likely future development of nuclear separation technologies.

In the Introduction, Jack Watson, Field Coordinator for the DOE Efficient Separations and Processing Program and an authority on separation research, provides an overview of the book and considers the necessity of public acceptance of nuclear-based technologies for future development of separation science. He points out that separations research and development can reduce the risks of nuclear operations and thus increase the willingness of the public to accept them, that nuclear energy is the only known energy source that can be increased significantly far into the future, and that the continuing risks of burning fossil fuels are greater than the risks of using nuclear energy. He also summarizes some of the essential concepts of actinide chemistry, since it is the basis for all nuclear separation processes.

Gregory Choppin, the author of Chapters I and IV, participated in the discovery of mendelevium, element 101, at the University of California, Berkeley. He is R. O. Lawton Distinguished Professor of Chemistry at Florida State University and an expert on actinide chemistry and separation science. He outlines the history and accomplishments of separation technology, especially in relation to spent reactor fuel processing. He points out that the era of large-scale nuclear applications began with a relatively simple precipitation process for separating plutonium from dissolved spent nuclear fuel and that the development of the PUREX separation process was a major improvement, which has become the basis of essentially all spent-fuel reprocessing. Separation technologies have played key roles in all major applications of nuclear science. Developing and building nuclear weapons required separation methods to remove plutonium from the irradiated fuel and to remove both "unburned" uranium and plutonium for re-use as fresh nuclear reactor fuel. In recent years, as the public has grown more concerned about the effects of radioactivity and as governments have begun to clean their aging nuclear facilities and sites, the need for separation methods to remove radioactive components from contaminated wastes, surfaces, soils, and ground water has become critical.

The author of Chapter II is Teresa Fryberger, Associate Deputy Assistant Secretary for Science and Technology within the Department of Energy's Environmental Management Program. She discusses the role of separation science and technology in the United States weapons complex cleanup. The challenge is enormous: fifty years of nuclear technology and weapons development have left the U.S. Department of Energy (DOE) with the task to clean 3,700 contaminated sites, more than a trillion gallons of radioactive and mixed wastes stored in 322 tanks, 3 million cubic meters of radioactive or hazardous buried wastes, 250 million cubic meters of contaminated soils, more than 600 billion gallons of contaminated ground water, and over 2,000 facilities that require decontamination and decommissioning at an estimated cost of $200 to $350 billion over 75 years. Separating the nuclear wastes rather than placing them in a single stable waste form is essential to give future generations more options for subsequent handling of these materials. The secondary wastes produced by cleanup operations also need to be carefully considered, since their treatment is likely to contribute significantly to the risks and costs of nuclear waste operations.

Boris Myasoedov, Academician and Director of the Laboratory of Radiochemistry of the Vernadsky Institute of the Russian Academy of Science, describes in the next chapter the role of separation science and technology in solving pressing environmental problems in Russia. A number of factors has made artificial radionuclides a persistent and very dangerous environmental problem in many regions of that country. Among these are nuclear weapons tests in the atmosphere and underground; nuclear power plants with nuclear fuel cycles producing and accumulating weapon-grade plutonium (Mayak Production Association in Ozersk, Siberian Chemical Plant in Seversk, Mining and Chemical Plant in Krasnoyarsk-26); operations of and accidents at nuclear power stations; and the unsanctioned disposal of radioactive contaminants from nuclear-powered ships and submarines and from nuclear waste sites into the oceans. Only in the 1960s did society begin to realize the global consequences of these activities, and attention shifted to maintaining sustainable development, including remediation of the polluted territories, study of radionuclide behavior in nature, reduction of the amount of radioactive wastes, and development of technology for long-term radioactive waste storage.

In Chapter IV, Gregory Choppin discusses the major advantages of non-aqueous processing (e.g., pyrometallurgical processing) of spent nuclear fuel. This approach will not necessarily achieve the same degree of purity from uranium and plutonium isotopes, but it can give sufficient purity for making new fuel. Non-aqueous processing can reduce waste volume considerably. The decreased purification may in fact be an advantage in safeguarding the recycled plutonium fuel because sufficient radioactivity would remain to make movement of the fuel easy to detect. The unsanctioned handling of the recycled fuels would also be hazardous without the large, shielded equipment used at reactors and fuel processing centers.

Valery Romanovsky, Deputy Director General of the Khlopin Radium Institute in St. Petersburg, Russia, details in Chapter V a successful example of international cooperation, the development of a universal extraction technology (the UNEX process) for processing and treating high-level wastes. This technology is being developed cooperatively at both Russian and U.S. institutions for applications in both countries. Since the number of separation scientists as well as funds for research are decreasing worldwide, international cooperation will become more and more essential if full use is to be made of remaining talent and facilities.

Donald Oakley, Associate Director of the Institute for International Cooperative Environmental Research and an expert on environmental and health effects of civilian and military nuclear activities, describes in the next chapter the long-term health risks. The major source of public exposure to radioactivity may result from mining and milling waste and from disposal or reprocessing of nuclear waste and not from the operation of nuclear plants or nuclear bomb tests as commonly perceived. Proposed improvements in waste partitioning, such as non-aqueous means for separations, may decrease those risks by significantly reducing the volume of secondary wastes that must be either treated or isolated to prevent public radiation exposure. As advances are made towards smaller, remotely controlled separation processes, occupational exposures will decline. Thus separation technologies can reduce the public health and environmental risks of nuclear power and thereby make it more acceptable to the public.

An option that would complement the currently accepted practice in the United States of vitrification and burial of nuclear waste is discussed in the last chapter by Hans Plendl, Emeritus Professor of Physics at Florida State University and an expert on accelerator-driven transmutation of nuclear waste. Neutrons from the fission process in nuclear reactors or from the accelerator-beam-induced spallation process can be used to transmute long-lived isotopes to shorter-lived ones in order to leave less or no long-term radioactive legacy. The use of reactor-produced neutrons may be less costly, but it would generate additional long-lived isotopes that would require disposal. The transmutation option would require extensive and challenging separation processes because of the great variety of nuclides produced by that method.

From the issues discussed in this volume and during the workshop on which it is based, the following recommendations for the future development of separation science and technology have emerged:

- New separation technologies are urgently needed to minimize nuclear wastes, including secondary wastes, in current and future nuclear operations. The partitioning of wastes to separate the long-lived radioisotopes and the production of special waste forms for safe disposal of these isotopes need to be improved.

- Separation and waste form experts need to interact and cooperate more closely since separations and waste form production are sequential operations.

- Environmental safety needs to be improved, and remediation of areas affected by past operations needs to continue.

- The principal issues inhibiting progress in separation science are more economic than technical. Thus the key to the development and acceptance of new separation processes is to reduce their cost.

- Transmutation and other alternatives to vitrification and burial need to be considered as complementary rather than competitive methods. The problem of long-term exposure is a critical issue: at the present time, the long-lived isotopes should be separated and stored, not buried, so that transmutation and other options can be considered in the future.

- Surplus weapons plutonium and other fissile materials need to be safely stored. Such materials should be considered as a potential resource, not a waste. Possibilities such as their safe use in the form of mixed fuel should be investigated further.

- The worldwide decrease in the number of faculty appointments for nuclear and radiochemistry needs to be reversed, since the resulting decrease in trained professionals has already become a serious impediment to future development of separation science and technology.

- Public acceptance of nuclear issues will play a strong role in decisions on the continuation of R&D and future uses of nuclear technologies. An international effort to attract public support for necessary nuclear operations and for long-term research toward reducing the risk of separations and nuclear waste operations are clearly needed. The existing legacy of accumulated nuclear waste materials will require separations even if production of nuclear energy were to cease altogether.

- Organizations that provide funding for research and development need to be made aware that the best available science and technology is not always being considered in current attempts to solve environmental problems caused by nuclear operations and to prevent future ones. For example, a policy of disposing all of our nuclear wastes in a single final waste form such as borosilicate glass may be too limited and closes other promising options.

- International cooperation is essential for the future development of separation science. Many of the separation technologies that are currently used or are being considered for future use in countries were initially developed in other countries. Nuclear waste management is an international problem, and we must avoid limiting ourselves by looking for strictly national solutions.

INTRODUCTION

Jack S. Watson
Oak Ridge National Laboratory
Oak Ridge, Tennessee 37831, U.S.A.

The nuclear era would not have been possible without development of the means to separate nuclear materials. Materials must be purified before they can be used in reactors and other nuclear devices. For many decades, developing and building nuclear weapons required separation methods to remove weapon-grade plutonium from the irradiated fuel and to remove "unburned" uranium and plutonium from spent fuel from reactors for reuse as fresh reactor fuel. Today, the most critical need is for separation methods to remove radioactive components from contaminated wastes, surfaces of working areas, soils, and ground water. This last need has become even more pressing as the public has grown more concerned with the hazards of radioactivity and as governments have begun to cleanup their aging nuclear facilities and sites.

In the past 50 to 60 years, nuclear science gradually became an industry rather than a laboratory pursuit. In the 1940s, 1950s, and even 1960s, nuclear science and technology was a glamour field that enjoyed the respect given in later decades to space technology, then to computer technology, and most recently to Internet technology. In the earlier decades, most people viewed nuclear energy as a solution to many of our energy needs and as a possible solution for some medical problems. Most people working in this field probably were shocked by the transition in the public's perception of nuclear energy from a hope for the future to a curse on the future.

1. Nuclear Energy and the Environment

The public view of nuclear energy has shifted as a result of highly publicized accidents at nuclear reactors and the arguments of anti-nuclear advocates. Although not directly related to peaceful uses of nuclear energy, the accumulation of huge stockpiles of nuclear bombs in several countries has probably also affected public opinion. For several decades, anti-nuclear sentiment has grown, affecting government funding for nuclear energy and playing a major role in promulgation of regulations and policies that have raised the cost of nuclear power substantially. The future development of nuclear separations may depend upon the future of nuclear energy, but the existing nuclear legacy will continue to require nuclear separations even if the production of nuclear energy were to decrease or stop altogether.

The risks of global warming are likely to be more serious than the risks from nuclear power. Substituting nuclear reactors for coal or other fossil-fuel-fired power plants will reduce carbon dioxide emissions responsible for global warming. The very rapid decline in natural gas inventories and the sharp increase in gas prices in the U.S. during the last quarter of 2000 also make nuclear power a more attractive option for the future. However, the solutions discussed for reducing greenhouse emissions usually include carbon sequestration, biomass utilization, conservation, and other solutions rather than the use of nuclear energy. Until the general public views the benefits of nuclear power to be greater than the environmental and safety risks, there will be little significant growth in the use of nuclear energy.

Nuclear power remains one of the few large energy sources that has not been fully exploited. If greenhouse effects do not curtail consumption of fossil fuels in the near future, the decreasing availability of fossil fuels will affect consumption in the long term due to increasing prices. Any shortage in fossil fuels, however, could be several decades away unless oil cartels artificially create a shortage. Conservation can compensate for some decrease in the use of fossil fuels, and renewable energy sources (solar, wind, biomass) can be used to a greater degree. The demand for energy will also be affected by population growth and by the standard of living of that population. Although an increase in the standard of living may be possible without an increase in energy consumption, in the past this usually has not occurred. When we do need major increases in energy supplies, nuclear energy is the one large source that can be increased very significantly, even far into the future. In many countries, including the U.S., nuclear reactors generate a significant portion of the electricity. New reactors continue to be built in other places in the world, but not in the more strongly anti-nuclear countries. Radioactive wastes are scattered through all countries that have had nuclear programs or that have used significant quantities of radioisotopes for medicine and other purposes.

The role of separation technologies in the nuclear industry begins with the preparation of raw materials, especially uranium, for producing nuclear reactor fuels. Ion exchange and solvent extraction (liquid-liquid extraction) methods have been developed and are deployed throughout the world where suitable uranium ores are found. During the last decade, the decreased demand for uranium from the weapons industries in the U.S. and Russia has made larger quantities of enriched uranium available for reactor use.

Spent nuclear fuel is not reprocessed presently in the U.S., but reprocessing continues elsewhere in the world. The present cost of uranium does not make reprocessing of spent fuel economical in the U.S. Also, there are serious concerns with safeguarding the plutonium recovered from the fuel to insure that it is recycled to power reactors and is not obtained by terrorists or rogue nations. Plutonium from power reactors does not make good weapons fuel because the higher "burn-up" of the fuel results in a high concentration of plutonium isotopes with masses greater than the more fissile 239 isotope, but even an inferior product can be a threat in the wrong hands. Assuming that the safeguard issues can be resolved and nuclear power grows, or at least does not decline greatly, spent fuel reprocessing is likely to resume in the U.S. when increases in the price of uranium provide an economic incentive. The present glut in the supply of uranium has kept down the cost of uranium and hence the need for the reprocessing of spent reactor fuel.

2. Evaluation of Health Risks of the Nuclear Fuel Cycle

In one of the chapters of this volume, Don Oakley discusses the long-term health risks of the nuclear fuel cycle and points out that the actual risks may be significantly different from the public's view of the risks (Chapter VI). For example, the major source of public exposure to radioactivity may be from mining and milling waste and disposal of spent fuel or reprocessing waste rather than from reactor accidents. Separation technologies can play important roles in reducing these risks. Proposed improvements in waste partitioning, such as non-aqueous means for separations, hold promise for reducing significantly the volume of secondary wastes that must be either treated or isolated to prevent public exposure to radiation. Improved separation methods will also be needed to support future technologies for effective disposition of nuclear waste. As advances are made toward smaller, remote-controlled separation processes, occupational exposures can be expected to decline. Application and further development of separation technologies can reduce the risks inherent in nuclear power generation and can provide solutions to environmental problems associated with that activity.

3. Nuclear Waste Reduction

Separation technologies are needed to reduce the cost of cleaning radioactive contaminants from stored wastes, contaminated soil, and ground water. Most radioactive materials have half-lives of not more than a few decades; so most radioactivity, unlike contamination from toxic non-radioactive metals, will dissipate with time. However, the time required is very long for some nuclides, most notably for the actinide elements and technetium, which have half-lives of thousands of years or more. These materials are essentially permanent contaminants, as they remain radioactive for millennia. The conventional approach in the U.S. to handle wastes containing long half-life radioactive contamination is to incorporate them into a very stable waste form and store them in a relatively dry, stable, and remote site. The currently favored waste form incorporates the radioactive material in a glass matrix (vitrified), and the planned disposal site is in Yucca Mountain at the Nevada Test Site. As long as spent fuel reprocessing does not occur in the U.S., spent reactor fuel elements will be sent to that repository.

Another option being considered in the U.S. and in other countries is to use neutrons to transmute the long-lived isotopes to shorter-lived isotopes. Neutrons required for the transmutation could come from suitable nuclear reactors or from accelerators. The use of reactors would probably be less costly, but reactors would also generate additional long-lived isotopes that would need to be transmuted as well. Transmutation operations require complex and challenging separation processes that could be as extensive and costly as the transmutation process. Hans Plendl discusses the transmutation of unwanted nuclear materials (Chapter VII).

The progress in improving separation technologies for nuclear materials has been remarkable over the past 50 years. But, problems and obstacles to future developments remain. Gregory Choppin describes the progress made in spent fuel reprocessing (Chapter I). The nuclear era began with a relatively crude precipitation process for recovering plutonium from dissolved spent fuel. That approach did not recover the uranium and left a residual waste enlarged by substantial quantities of precipitating agents. The development of a solvent extraction process that recovered and separated the uranium as well as the plutonium was a major improvement. This process, however, required the addition of high concentrations of sodium nitrate to improve the extraction of the uranium and plutonium, and resulted in an additional mass of solid materials in the high-level waste produced. The development of the PUREX process was a further major improvement and is used in all spent fuel processing facilities operating now or that have operated during the past few decades. A major advantage of the PUREX process is use of nitric acid rather than sodium nitrate to aid in the extraction of uranium and plutonium.

Gregory Choppin also considers the likely further evolution of spent fuel reprocessing

(Chapter IV). He sees major advantages for non-aqueous processing (pyrometallurgical processing, that is, processing with molten metals and/or molten salts). This approach will not necessarily achieve the same degree of purity of the uranium and plutonium products as aqueous processes, but it can yield purities sufficient for making new fuel. Non-aqueous processing can reduce waste volume considerably, and the modest purification can be an advantage in safeguarding the recycled plutonium fuel, because sufficient radioactivity will remain in the recycled fuel, making movement of the fuel easy to detect. Handling the less pure recycled fuels would be hazardous without large, shielded equipment available at reactors and fuel processing centers, and thus concealing the fuel would be even more difficult.

New research throughout the world has improved separation methods for handling the residual radioactivity both from current nuclear operations and from the legacy of past operations. Teresa Fryberger points out that several separation approaches are likely to be needed to handle nuclear wastes most effectively (Chapter II). Costs for handling nuclear wastes, especially the high-level wastes, can be extremely high, and seeking optimal systems to minimize the costs will be important. One process will not be optimal for all wastes, not even for a single type of waste (high-level, low-level, etc.). Separation of wastes, rather than placing them all in a single waste form, will leave future generations more options for subsequent handling and disposal of these materials. Careful consideration should also be given to all the secondary wastes produced by cleanup operations. Treatment of wastes could contribute significantly to the risks and costs of nuclear waste operations.

Boris Myasoedov discusses the environmental impacts of nuclear activities in Russia. He describes also several new solid sorbents (adsorbents or ion exchange materials) and extraction agents that may potentially be used in transmutation operations to separate key materials like transuranium (actinide) elements from stored high-level wastes (Chapter III).

4. International Cooperation

Valery Romanovsky describes important results from the Khlopin Radium Institute (St. Petersburg, Russia) with special mention of the universal extraction process UNEX that has been developed there in cooperation with the Idaho National Engineering and Environmental Laboratory (INEEL) (Idaho Falls, Idaho, U.S.A.). This process makes it possible for the first time to simultaneously separate and recover Cs, Sr, and actinide and rare-earth elements. Processes in use up to now were able to simultaneously separate and recover only some of these elements in a single process step. This process is being tested at INEEL and is expected to be applied to high-level liquid waste both in the U.S. and in Russia (Chapter V).

As the author points out, this international cooperation can serve as an example for future cooperation. The limited expertise and facilities available internationally need to be used cooperatively to solve the nuclear separation problems of the future. The loss in capabilities in the separation of radioactive materials extends well beyond the U.S. Scientists in Russia, the Czech Republic, and other countries have also noted problems resulting both from the aging of the trained workers in separations R&D and from a lack of replacements due to a decrease in faculty positions in this field.

5. Chemistry of the Actinide Elements

Separation of radioactive materials involves a number of chemical elements, and some of these, the actinides, are essentially never used in any other applications, so that the chemistry of these elements is not widely known. This group, from actinium (No. 89) to lawrencium (No. 103), contains the reactor fuels thorium (Th), uranium (U), and plutonium (Pu) as well as the so-called minor actinides neptunium (Np), americium (Am), and curium (Cm) produced by neutron capture in reactor operation. Actinide chemistry is, therefore, at the heart of many key aspects of radioactive element separations. One must understand actinide chemistry to make significant improvements in methods for removing and separating these long-term radioactive isotopes that present major long-term hazards to the environment and to humans. Better understanding of the behavior of actinides in the environment and in the human body is also needed for more rational and more effective regulatory rules and constraints.

The actinide elements have both unusual and interesting chemistry. Their location on the Periodic Table right below the lanthanide group suggests a chemistry much like these elements, and that similarity holds true to a great extent. However, the similarity to the lanthanides and even the similarities among the actinides are limited. The actinides often have several valence states, and the valence states of each actinide element differ in the same identical chemical environments.

The different valence states of different actinide elements can result in quite different chemical properties, including different quantitative properties. For example, in sea water [1] americium and curium exist only in the trivalent state. Uranium, however, can exist in either the hexavalent (VI) state (the uranyl ion or UO_2^{++}) or in the tetravalent state (IV). The higher oxidation state, uranyl ion, is more likely to be found near the soil surface where the environment is usually more oxidative. Neptunium is highly likely to be pentavalent (V), and plutonium can be in any oxidation state from the tetravalent to the hexavalent. Each oxidation state has a different chemistry, and significant differences in the chemistry of each actinide element exist even in the same oxidation state. The accessibility of multiple oxidation states of the various actinides and the possible presence of several of these states simultaneously for some actinides complicate the modeling of their behavior.

The importance of the oxidation state of actinide ions is illustrated by their behavior in common soil conditions. The lower valence actinides are more likely to be insoluble and migrate very slowly, if at all, in ground water. Uranium (IV) has a very low solubility in most ground waters and can usually be considered non-mobile. Uranium (VI), on the other hand, can be more soluble. This strong tendency of some valence states of the actinides to form complexes makes the chemistry and behavior of the actinides strongly dependent on the chemical environment. The strong interactions with strong-base anions play an important role in the behavior of the actinides in the environment, and multiple bonding plays a major role in the development of selective extraction reagents to remove and separate actinides. The uranium (VI) ion can interact strongly with anions present in ground water, and those interactions affect its behavior. For instance, carbonate ions readily complex with uranyl ions, and the presence of significant concentration of carbonate ions can result in much of the uranium being highly bonded with anions, and the resulting complex can be in an anionic form. These complexes are important to understanding the migration of uranium in the environment.

Actinide cations are "hard acid" cations whose bonding is strongly ionic. These actinide cations bond with "hard base" molecules such as those containing oxygen and fluorine donors (including water molecules). In aqueous solutions, there can also be complex bonding to softer donors such as nitrogen and sulfur atoms, but these bonds are likely to be weaker and of importance only when involved with multiple bonding ligands (multidentate ligands with several bonding sites). The bond strength is proportional to the effective charge density of the metal, and the order among the oxidation states is:

$$AnO_2^+ < An^{3+} < AnO_2^{2+} < An^{4+}.$$

This trend agrees with the experimental evaluation of the effective charge densities of NpO_2^+ and UO_2^{2+} as +2.2 and +3.3, respectively. The electrostatic interactions plus steric interference influence the coordination number about the actinide, which can vary from 6 to 12 for simple actinide complexes and from 2 to 8 for actinyl (dioxo) compounds such as UO_2^{2+}.

Plutonium chemistry can be even more complex [2]. There is a greater potential for multiple oxidation states to exist at one time, and plutonium may not even behave as a single element, at least if the rate of transition from one oxidation state to the other is not rapid. Also, soluble plutonium (V) can exist in equilibrium with insoluble plutonium (IV).

The rate of change of actinides from one oxidation state to another can also be important, especially where ground water flows from one environment to another. The rate of change appears to be catalyzed by some soluble or insoluble minerals present. The presence of ligands that can attach to the actinide ions can also affect the rate of oxidation or reduction.

Since the actinides are among the key elements of most concern and interest in radioactive contamination, complex computer codes have been developed to predict the behavior of these materials in a variety of environmental conditions. But these codes often have to rely upon questionable data or on estimates when no data at all are available. In recent years, little work has been done to produce new data for such models. In fact, only a few laboratories in the world remain with personnel, facilities, and capabilities to obtain the needed data. International cooperation in this area will therefore be essential and will yield significant benefits.

6. Conclusion

Although the following chapters cannot cover all of the chemical and technological aspects of separation science, they do provide valuable assessments by key people who have worked in this field for many years. Broader issues associated with nuclear power production and other nuclear activities in the U.S. and throughout the world are also discussed.

References

1. Choppin, G. R., and Wong, P. L. (1998) The Chemistry of Actinide Behavior in Marine Systems, *Aquatic Geochemistry* **4**, 77.
2. Choppin, G. R., and Stout, B. E. (Dec. 1991) Plutonium – the Element of Surprise, *Chemistry in Britain*, 1126.

 **Jack S. Watson** has more than 40 years of experience at the Oak Ridge National Laboratory, mostly on some form of separations research. In the past fifteen years, most of his work has been concerned with environmental and waste issues. He is currently the Field Coordinator for the Department of Energy's Environmental Management Efficient Separations and Processing Crosscutting Program. He is a Fellow of the American Institute of Chemical Engineers, and he has organized numerous conferences, most notably many of the biennial Symposia on Separation Science and Technology for Energy Applications. He is an Adjunct Professor of Chemical Engineering at the University of Tennessee and an Associate Editor of the journal *Separation Science and Technology*.

Chemical Separations
in Nuclear Waste Management

The State of the Art and a Look to the Future

PART 1

NUCLEAR SEPARATION TECHNOLOGIES

State of the Art

CHAPTER **I**

OVERVIEW OF CHEMICAL SEPARATION METHODS

GREGORY R. CHOPPIN
Department of Chemistry
Florida State University
Tallahassee, Florida 32306, U.S.A.

ABSTRACT

Chemical separation science and technologies are presented from a historical perspective. Separations involving radioactive actinide elements began with the discovery of radioactivity. All early separation methods involved precipitation. In the Manhattan Project in the U.S.A. during World War II, plutonium was isolated initially by co-precipitation with bismuth phosphate in order to extract it from spent uranium fuel elements. This method was soon replaced by the development of non-precipitation techniques involving solvent extraction. In one such technique, the PUREX (plutonium and uranium recovery by extraction) process, tributyl phosphate was used to extract the nitrates of uranium and plutonium from the fission products. It has remained the most commonly used separation system for processing irradiated fuel elements to the present time. PUREX and other traditional separation systems are reviewed to illustrate the characteristic features of separation methods. Various other methods are also considered with emphasis on their limitations and advantages.

Ion-exchange processes for the separation of actinide elements from fission products and the resins, both anionic and cationic, that have been developed in recent years specifically for this purpose are also considered. In addition to the conventional solvent extraction and ion-exchange separations, more innovative systems under study involving membrane separations or the use of natural agents (such as sidereophores) for treatment of contaminated equipment and for environmental remediation are reviewed.

1. Earliest Separations

Separations of radioactive elements began with the discovery of radioactivity. In 1896 when W. Crookes and H. Becquerel added carbonate anion to a solution containing uranium, the uranium remained in the supernate as a soluble uranyl carbonate complex, and a precipitate containing b and g radioactivity formed. Marie and Pierre Curie separated components of pitchblende to isolate the non-uranium radioactive elements, and in 1898 announced the discovery of the new element polonium whose radioactivity is 400 times greater than that of uranium. Between 1932 and 1939, approximately fifty research papers had described the discovery and study of what were believed to be transuranium elements with Z = 93, 94, 95, and 96. In 1939, after conducting very careful separations on irradiated uranium samples, O. Hahn and F. Strassmann concluded that these elements were, in fact, elements with atomic numbers below 60, and they thereby discovered the nuclear fission process. This led to new experiments in 1940 in which neptunium (Z = 93) and plutonium (Z = 94) were isolated and identified using a redox (oxidation-reduction process) cycle with bromate as the oxidizing agent followed by precipitation in which the reduced cations were co-precipitated with crystalline LaF_3. Both the redox and fluoride precipitation separations are still standard procedures in actinide separation science.

Precipitation remained the predominant separation technique in radiochemistry until the development of the atomic bomb during the Manhattan Project. The scientists working on this project were the first to separate plutonium from uranium and its decay and fission products using kilogram-scale radiochemical separations. The first large-scale separation was the Bismuth Phosphate process, which used $BiPO_4$ as carrier for the insoluble phosphates of Pu(III) and Pu(IV) (Figure 1-1) [1]. Advantage was taken of the relative stability of the oxidation states of uranium (VI) and of most fission products, compared to the redox lability of plutonium (III, IV, V, and VI). In the early processing work, plutonium was isolated by precipitation in the reduced state as PuF_3 or PuF_4,

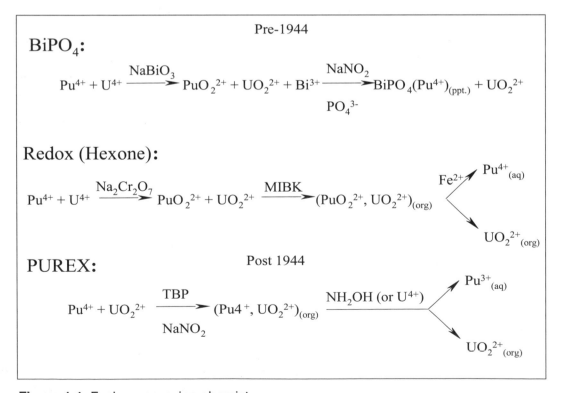

Figure 1-1. Fuel reprocessing chemistry.

together with all insoluble fission product (FP) fluorides. This was followed by a second stage, which involves dissolving the precipitate, then oxidizing the plutonium to the VI state, followed by another fluoride precipitation, leaving relatively pure plutonium in the supernatant solution. In a final step plutonium was again reduced and precipitated as fluoride. This principle was used for the first isolation of hundreds of kilograms of plutonium at the Hanford Engineering Works, U.S.A., with phosphate precipitation of Pu^{3+} and Pu^{4+}, but not of PuO_2^{2+}, which does not form an insoluble phosphate.

This principle of oxidizing and reducing plutonium at various stages of the purification scheme has been retained in subsequent processes. No other element has the same set of redox *and* chemical properties as plutonium, though some elements behave like Pu^{3+} (e.g., the lanthanides), some like Pu^{4+} (e.g., zirconium), and some like PuO^+ (e.g., uranium). Numerous redox agents have been used, such as $K_2Cr_2O_7$, $NaNO_2$, hydrazine, and ferrous sulfamate.

A major disadvantage of the Bismuth Phosphate process is that it requires batch operation, which recovers plutonium but not uranium (which remains in the waste stream). It was soon replaced by solvent extraction processes that occur in continuous, remote operations and that isolate both uranium and plutonium from the fission products. The first solvent extraction system adopted on a large-scale used methyl isobutyl ketone (MIBK or hexone) in the redox process (Figure 1-1) [2]. Hexone acts as an extracting solvent, forming adduct compounds with coordinatively unsaturated actinide nitrates, as expressed in:

$$Pu^{4+} + 4NO_3^- + 2S(org) \times Pu(NO_3)_4S_2(org)$$

where S= MIBK. While these neutral hexone compounds of the actinides are extracted into the organic phase, essentially all of the radioactive fission products remain in the aqueous waste stream. The major disadvantage of the redox process is the tendency of hexone to decompose slowly in contact with the strong nitric acid used as the aqueous phase.

Further research on solvent extraction systems led to the use of tributyl phosphate (TBP) as an extractant to isolate uranium and plutonium from fission products. TBP has the advantages of being more stable, less flammable, and resulting in better separation than hexone. The PUREX process, using a solution of 30% TBP in kerosene as the organic phase, has been the most employed of all large-scale radiochemical separation techniques and remains in common use today [3].

In the PUREX process (Figures 1-1 and 1-2) after the irradiated uranium is dissolved in 6-8 M nitric acid solution, a small amount of nitrite is added to stabilize the plutonium in the tetravalent state, and the solution is contacted with an organic phase of TBP dissolved in kerosene [4]. Only uranium (as UO_2^{2+}), plutonium (as Pu^{4+}), and neptunium (as NpO_2^{2+}) are extracted from the aqueous nitrate feed solution. The tetravalent plutonium is partitioned from the hexavalent uranium by reducing Pu(IV) to Pu(III), which allows the plutonium to be back-extracted into a 6M nitric acid aqueous phase, while the UO_2^{+2} remains in the organic solvent phase in dilute nitric acid.

The PUREX process depends on the relative ease of redox of plutonium, which allows recovery of purified plutonium from the much greater concentration of uranium. PUREX does not require inorganic salts for efficient extraction, and both solvent and aqueous reagent streams can be recycled for reuse in the process, thereby reducing the amount of the waste from the process. The fission product fractions can be stored for treatment for final disposal. The plutonium and uranium are produced from fission products separately with decontamination factors of greater than 10^7 for both actinides. The amount of

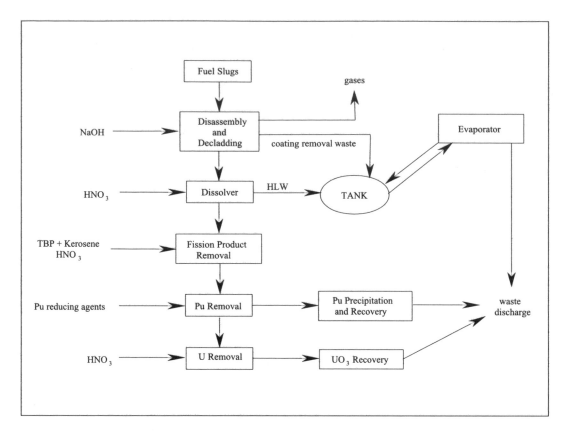

Figure 1-2. Flowsheet for PUREX process.

plutonium and uranium that remains with the fission products in the aqueous waste streams may range from a few tenths to several percent.

PUREX remains the generally accepted industrial method of separation resulting in high quantity production of uranium and plutonium from fission products. This process could be adapted to treat defense wastes from production of plutonium by concentrating the actinide elements into fractions of relatively small volume for storage or for destruction by subsequent fissioning in reactors or accelerators (transmutation). However, depending on the composition of the waste, additional separation steps may have to be added to the PUREX process to obtain better separations of certain elements. For example, it is possible to selectively remove long-lived (e.g., ^{99}Tc) or highly radioactive (e.g., ^{137}Cs) fission products from the waste streams using ion-exchange processes [5,6]. One example of an additional solvent extraction system is the SREX process, which uses a crown ether to isolate Sr from the PUREX waste streams [7].

2. Present Directions in Separations

Present research efforts are focusing on safe and cost-effective management of the radioactive wastes produced in nuclear weapons programs (defense wastes) and spent fuel from operation of nuclear power plants (commercial wastes).

Defense wastes originate from the reprocessing of irradiated uranium for the production of plutonium and commonly contain a mixture of fission products and transuranic elements and their decay products, as well as radionuclides from neutron activation. In the United States, these wastes are presently stored in underground tanks at various DOE facilities. One hundred seventy-seven of the tanks are located at the Hanford site [8]. The current policy for remediation for these wastes is to isolate and concentrate the longer-lived and more toxic radioactive species. This material, defined as high-level waste (HLW), is extremely costly to handle and store, making minimization of its volume a valuable objective. The residues from the separations are less radioactive, defined as low-level waste (LLW) (Table 1-1), but much larger in volume. Such LLW is stored commonly in surface or near-surface repositories.

TABLE 1-1. Nuclear Waste Types

High-Level Waste (HLW): highly radioactive waste resulting from chemical processing of spent nuclear fuel and irradiated target assemblies; usually a combination of TRUs and fission products.

Transuranium (TRU) Waste: waste that contains a-emitting TRU elements with a $t_{1/2}$ greater than 20 years and a total activity of more than 100 nanocuries per gram of waste; the great majority results from weapons production processes and contains plutonium.

Low-Level Waste (LLW): radioactive waste not classified as HLW, TRU, spent nuclear fuel, or byproduct material and acceptable for disposal in a licensed land disposal facility; typically includes discarded radioactive materials such as rags, construction rubble, and glass that are only slightly or moderately contaminated.

The radionuclides in commercial waste (spent reactor fuel elements) differ significantly from those in defense waste. The main steps in processing spent nuclear fuel are shown in Figure 1-3. The HLW from nuclear power plants consists of a spectrum of radionuclides (Table1-2). In defense wastes the radionuclides of major concern usually are ^{137}Cs, ^{90}Sr, ^{239}Pu, and ^{241}Am. ^{99}Tc and ^{237}Np become the more important radionuclides at longer storage times. A major concern in separating HLW is not the radionuclides but the chemicals added during processing. The U.S. defense wastes were neutralized for storage and include inorganic compounds such as $NaNO_3$, $NaNO_2$, and lesser amounts of caustic $Fe(OH)_3$ and organic compounds. Conventional processing and chemical engineering techniques can be used with some modification for remote handling and for concentrating and isolating many of these bulk materials. However, more sophisticated techniques are required for separating and isolating the radionuclides. For example, selective dissolution and dilution may remove many bulk components from less soluble residues. For the radioactive residues after dissolution, solvent extraction may be used to isolate transuranium elements; ion exchange, for removing Sr, Cs, and Tc; and flocculation, for removing suspended tracer-level plutonium species.

3. Aqueous Processes

Aqueous processes have been the dominant means of actinide separation and are likely to continue to play that role in the treatment of spent fuel in the near future. These separations depend on the differences in the chemical properties of the dissolved species, such as their reduction potentials, their complexation strength with various ligands and extractants, their affinity to ion-exchange resins, and their transport behavior through mem-

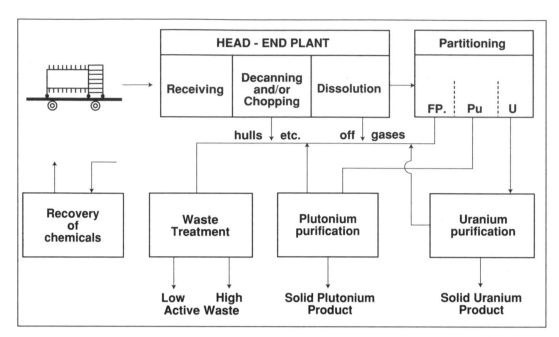

Figure 1-3. Main steps in reprocessing spent nuclear fuels.

TABLE 1-2. Composition of HLW from PUREX Reprocessing of 1 t Light Water Reactor Fuel with a Burnup of 33 000 MW d/t and a Cooling Time of 3 years

Component	Weight (kg) in waste volume ($\sim 5m^3$)	Component	Weight (kg) in waste volume ($\sim 5m^3$)
Fission Products:		Corrosion Products:	
Group I (Rb, Cs)	2.94	Fe	1.1
Group II (Sr, Ba)	2.37	Cr	0.2
Group III (Y, Ln)	10.31	Ni	0.1
Zr	3.54	Actinides:	
Mo	3.32	U ($\sim 0.5\%$)	4.8
Tc	0.77	Np ($\sim 100\%$)	~ 0.44
Group VII (Ru, Rh, Pd)	4.02	Pu ($\sim 0.2\%$)	~ 0.018
Te	0.48	Am ($\sim 100\%$)	~ 0.28
Others	0.35	Cm ($\sim 100\%$)	~ 0.017
		Neutron poison: e.g., Gd	12

branes or in an electric field. The PUREX process can be easily adapted to the treatment of actinide containing wastes. Although it does not achieve enough decontamination to enable disposition of the waste streams as low-level waste, it does provide sufficiently clean feed streams for subsequent purification processes such as TRUEX [8].

The TRUEX (transuranic extraction) process, developed at Argonne National Laboratory (ANL), is based on the use of octyl-(phenyl)-*N,N*-disobutylcarbamoylmethyl-phosphine oxide (CMPO), dissolved in a liquid alkane (Figure 1-4). TRUEX treatment of

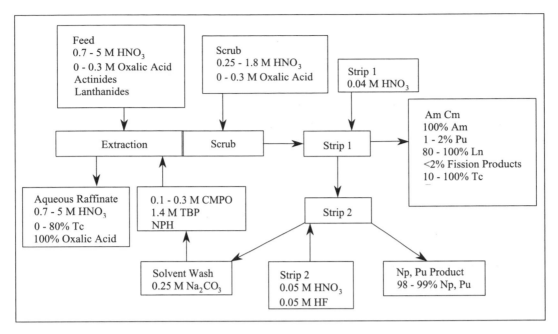

Figure 1-4. Flowsheet for TRUEX process.

the waste streams from PUREX treatment of irradiated uranium reduces the concentrations of residual U and Pu in the wastes by an additional factor of 10^2 to 10^3. However, TRUEX does not separate the tri- and tetravalent actinides from the lanthanide fission products.

A major drawback of the TRUEX process is that it requires aggressive stripping conditions to remove the tetra- and hexavalent transuranic elements from the organic phase. Because of the strong complexation strength of these metals by CMPO, conventional stripping with dilute acid does not work, and more powerful stripping agents must be used. Among the stripping agents being investigated are thermally unstable alkyl diphosphonic acids (TUCS reagents) [9]. These compounds form strong complexes with actinides in acidic solutions and decompose relatively easily at elevated temperatures. However, their potential use as agents for stripping into strongly acidic solutions of the residual transuranics is uncertain, because a decomposition product, phosphate, is an undesirable component in the waste stream and may interfere in the final waste vitrification process. Another disadvantage is the relatively high cost of CMPO, which could be reduced by large-scale production.

A considerable research effort has focused on the separation of trivalent actinides from trivalent lanthanides. Some lanthanide isotopes have high cross sections for neutron absorption and must be removed from recycled spent fuel as well as from the target material that is to be used for transmutation of actinides. The Trivalent Actinide Lanthanide Separation by Phosphorus Extractants and Aqueous Complexes (TALSPEAK) is based on the extraction of trivalent lanthanides with (2-ethylhexyl) phosphoric acid (HDEHP) from an aqueous solution of both lactic and diethylenetriamine pentaacetic acid (DTPA) at pH 2.5 to 3.0 [10]. Trivalent actinides remain in the aqueous phase. In the "Reverse TALS-PEAK" process, both actinides and lanthanides are extracted with HDEHP, with subsequent stripping of the actinides into an aqueous phase containing lactic acid and DTPA [11]. The TRAMEX process utilizes liquid cation exchangers like trialkylamines or tetraalkylammonium salts to selectively separate actinides from trivalent lanthanides and fission products [12] (Figure 1-5). Extractions are typically carried out from solutions of 10M LiCl, which requires a conversion step for the waste solutions as they generally have high nitrate concentrations (from the PUREX process). The highly corrosive properties of the strong chloride medium make this process rather unattractive for use on an industrial scale. If it were feasible to use the TRAMEX process in nitrate rather than chloride solutions, it would become a more attractive option.

4. Other Extractants

The solvent extraction processes discussed in the previous sections are either already well developed and applied by industry (e.g., PUREX) or well studied in the laboratory and ready for pilot-plant tests (e.g., TRUEX). To isolate a particular radioactive element from nuclear wastes, new extractants are being synthesized and studied in many

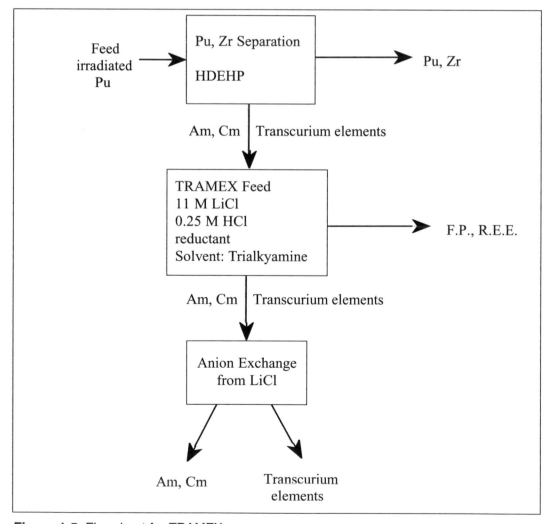

Figure 1-5. Flowsheet for TRAMEX process.

laboratories in the U.S. and in other countries. These include bidentate extractants such as diphosphine dioxides and diamides, stereospecific complexants such as crown ethers and sidereophores, "soft donor" complexants with active S or N atoms, and cobalt dicarbollide extractant.

Bidentate extractants: In addition to the carbamoylphosphine oxides used in the TRUEX process, other bidentate extractants, primarily diamides [13] and diphosphonic acids [14], are under consideration for actinide separations. These, generally, are strong extractants and have good radiation stability, but like CMPO extractants, they do not provide sufficient selectivity for the separation of actinides from lanthanides.

The bifunctional amide N,N-dimethyl-N,N-dibutyltetradecylmalonamide (DMDBT-DMA) dissolved in an aliphatic diluent was chosen as the extractant in the French DIAMEX process [15]. The major advantages of this extractant are its complete destruction by incineration as well as the ease of back-extraction of the actinides. Compared with extraction with CMPO, the extraction of Am(III) with DMDBTDMA requires higher nitric acid concentrations (>3M) and is relatively sensitive to the concentration of nitric acid below 4M. These are possible problems in its use in plant-scale operations.

Stereospecific extractants: Many of the extractants discussed so far show very strong complexation properties, but their selectivity is often insufficient for the separation of different radionuclides. Crown ethers, cryptands, and similar macrocyclic ligands can provide the desired high stereoselectivity for specific nuclides that match their cavity size and shape. Processes using the selectivity of crown ethers to remove Sr and Cs from HLW have been developed at Argonne National Laboratory [16]. ^{90}Sr and ^{137}Cs are the major sources of radioactivity and heat in defense wastes and their removal prior to further separation greatly simplifies subsequent waste handling and storage.

The Sr extraction/recovery process SREX uses a solution of 4,4'(5')-di-t-butylcyclo-hexano-18-crown-6 (DtBuCH18C6) and TBP in Isopar-L to extract ^{90}Sr from acidic waste [16]. The process uses a crown ether that has high specificity for extraction of ^{90}Sr. SREX can be used also to remove strontium from the waste streams of the PUREX process. Disadvantages of SREX include the solubility of the crown ether in the aqueous phase and its chemical and radiolytic degradation. Another concern of the SREX process is the cost of the reagent. Like the CMPO in the TRUEX process, the crown ether has to be manufactured on an industrial scale for the large-scale processing of nuclear wastes. The present synthetic process makes SREX much more expensive than classic commercial extractants such as TBP. This economic aspect is aggravated by loss of the extractant in the process due to its chemical and radiolytic degradation, its solubility in the aqueous phase, and third phase emulsification. Consequently, the extent of regeneration of the extractant is an important criterion that might determine the applicability of the SREX process for the large-scale treatment of nuclear wastes. A modification of the process, called CSEX-SREX, facilitates the removal of both Cs and Sr by extraction with a dibenzo-18-crown-6 derivative of proprietary composition [16]. Both processes achieve very high decontamination factors for Sr and Cs.

Soft donor complexants: Good separation factors have been reported between trivalent lanthanides and trivalent actinides for solvent extraction systems based on complexants with "soft" donor groups (e.g., nitrogen and sulfur). These complexants are based on amide functional groups [17] or on sulfur based b-diketone extractants [18]. However, techniques using oxidation/reduction steps can separate the actinides, including thorium and uranium, more simply and, usually, more efficiently. Research on soft donor ligand systems for separation of the transplutonium actinides from the lanthanides continues, and such a process could be used if an inexpensive ligand with very high radiation stability and low aqueous solubility is found.

Cobalt dicarbollide extractant: Dicarbollide (bisdicarbollycobaltate) [(p-(3)-1,2-$C_2B_9H_{11})_2$ Co] $^-$ has been shown to have very high selectivity and efficiency for extraction of Cs^{+1} and Sr^{+2} from 2 and 3 M HNO_3 solutions [19,20]. The hexacholoro and bromo derivatives of dicarbollide have good chemical and radiation stability. Dicarbollide is not as efficient for the extraction of trivalent actinides as it is for strontium and cesium, but useful extractions have been demonstrated from 0.3 to 0.5 M HNO_3 solutions. Useful extraction of cesium, strontium, barium, and americium has been achieved from PUREX wastes in which the acidity has been decreased using a solution of dicarbollide dissolved in p-nonylphenol-nonaethyleneoxide (a polyethyleneglycol). The process has been successfully applied to the treatment of 320 m^3 of actual HLW in Russia [21]. Coextraction of trivalent actinides is possible but needs to be improved. The hexachloro derivative of dicarbollide shows satisfying radiation stability, and dicarbollides are commercially available.

Solvent extraction in alkaline media: The separation processes discussed so far are designed mainly to treat acidic waste solutions. At some sites, the waste solutions presently stored in underground storage tanks have been neutralized by the addition of caustic substances to prevent corrosion of the carbon steel tanks. Thus, effective separation procedures for the treatment of these alkaline wastes are very unlikely to acidify the waste solutions before processing as this would greatly increase the volume and cannot be performed in the carbon steel tanks.

Several extractants have been studied that could remove actinides directly from alkaline solutions in the presence of complexing ligands. The latter would serve to prevent hydrolysis and precipitation of the metal ions. Russian scientists demonstrated the use of alkylpyrocatechols such as DOP 4-(a, a-dioctylethyl) pyrocatechol and amines like aliquat-336 to extract actinides from alkaline solutions containing carboxylic acids or aminocarboxylic acids [22]. The extraction of Pu(IV) from tartrate containing alkaline solutions with 20% Aliquat-336 in xylene has also been reported [23].

5. Ion-Exchange and Adsorption Processes

Ion-exchange has been used for decades to separate individual lanthanide and transuranic elements. Highly selective organic ion-exchange resins have been developed that provide excellent separation of particular radionuclides. A number of types of organic ion exchangers of high capacity and stability are being studied in many countries. Advanced inorganic ion exchangers are also under investigation, as they have higher radiation stability than the organic resins [24].

A promising new organic ion-exchange resin called Diphonix has been developed using a substituted diphosphonic acid resin as a strong complexing agent [25]. It removes actinides from strong acid solutions such as 10M nitric acid solution. The resin has been used successfully to remove toxic heavy metals from waste water and is commercially available.

An organic anion-exchange resin that shows remarkable radiation and chemical stability is the Reillex™ HPQ resin, a copolymer of 1-methyl-4-vinylpyridine and divinylbenzcne [26]. It is mainly used for selective plutonium isolation from 7-8 M nitric acid solutions. Only a few other metals form anionic complexes under these conditions.

Czech scientists have synthesized inorganic-organic composite absorbers that consist of finely divided inorganic ion exchangers incorporated into a binding matrix of modified polyacrylonitrile (PAN) [27]. These resins combine the high radiation stability of inorganic sorbents with the more favorable mechanical and granular properties of PAN. Various PAN-based resins have been successfully used in the treatment of radioactive waste, mainly for the removal of radiocesium. PAN-based resins can also be applied to moderately alkaline solutions with basicities up to 0.01 M NaOH.

The successful application of an inorganic sorbent on a pilot-industrial scale has been demonstrated using ferrocyanide sorbents for the removal of cesium [21]. Binding of Cs^+ occurs through substitution of potassium in the ferrocyanide sorbent ($K_{1.0}ME(II)_{1.5}Fe(II)$ $(CN)_6CnSiO_2$ with Me(II) = Ni, Cu or Zn). Other inorganic ion-exchange materials that are under consideration for the treatment of nuclear wastes include titanium phosphate, zeolites, silicotitanates, and clays. Their high capacities for some radionuclides and their high radiation stability also make them interesting candidates as matrices for final waste storage.

Silicotitanate materials can remove cesium from radioactive waste solutions in the presence of very high sodium concentrations [6]. Such specificity indicates the potential for these materials to remove ^{137}Cs from wastes contained in storage tanks in concentrated salt solutions. The silicotitanates also can remove Sr and actinides, but the selectivity of Sr over Ca is not as high as for Cs over Na. For actinides, the capacity of the silicotitanates is not high, but these materials may find use in caustic waste treatment where actinide solubility is low. These silicotitanates, as well as the zeolites, after sorption of the radionuclides, can be transformed into ceramic materials, which may be a useful form for direct permanent disposal after sufficient development and testing for long-term retention of the radionuclides.

6. Other Possible Separation Methods

Membrane processes: Membrane processes are widely used for industrial separations and may have applications in the treatment of nuclear wastes. An important criterion for the applicability of a membrane process to radioactive separations is the radiation resistance of the membranes. The defense wastes stored at the various DOE sites in the United States contain not only radioactive materials, but, usually, high concentrations of salts (e.g., sodium nitrate and nitrite), as well as other nonradioactive chemical reagents. The presence of these nonradioactive ingredients is a complicating factor in treatment of the nuclear wastes. Membrane processes such as dialysis, ultrafiltration, and facilitated transport have been proposed as techniques to separate the bulk salt from the radioactive components. Application of an electric field across a membrane allows monovalent ions to pass the membrane, leaving multivalent ions (e.g., actinides) behind. Separation of sodium nitrate and nitrite from the radioactive components would result in a significant reduction of waste volume. Membranes that are selective for monovalent ions are used extensively for the recovery of sodium chloride from seawater and are commercially available.

Researchers at the Plutonium Facility at Los Alamos National Laboratory have studied the separation of actinides from waste waters using water-soluble metal binding polymers [26]. These polymers selectively complex actinide ions and can be separated from the solution phase using ultrafiltration technology. Among the polymers under investigation are phosphonic acid derivatives of polyethylenimine.

Natural agents: A number of natural agents have been studied as inexpensive and potentially highly selective agents for the removal of actinides from aqueous solutions. Japanese investigators have demonstrated the ability of the tannin biomolecule to sorb uranium from seawater [28,29]. Uranium binding takes place via multiple adjacent hydroxyl groups, with a high selectivity for UO_2^{2+} over simple bivalent metal ions. Tannin can be immobilized on a variety of matrices to prevent leaching into the aqueous phase, the adsorption rates are rapid, and the sorbent can be used in both column and batch systems. Uranium desorption is easily achieved in dilute acid, allowing the use of the sorbent in repeated adsorption-desorption cycles.

Sidereophores (microbially produced chelating agents) have been found to form stable complexes of Pu(IV), which might be used to sequester plutonium from waste streams [30]. Other natural agents that have been considered for actinide separations are jimson weed to remove plutonium from sludges via binding to cell walls and a derivative of chitin that can be used in the form of porous beads to separate heavy metals from aqueous solutions [31].

Magnetic separation: Differences in magnetic susceptibility have been explored in the laboratory to separate uranium and plutonium from particulate wastes using the paramagnetism of the actinides [32,33]. In principle, the magnetic separation technique can be considered for the separation of any paramagnetic metal from diamagnetic or nonmagnetic materials. This method may find use in processing special wastes such as pyrochemical salts and incinerator ash although its employment for large-scale separations seems unlikely.

Electromigration (electrophoresis): In this technique, an electric field is applied to separate charged species from one another due to differences in mobilities. This technique has difficulties in scale up since heat dissipation is required to avoid convection currents that would perturb the migration pattern. Development of this technique for large-scale treatment of nuclear wastes seems unlikely.

7. Conclusion

Aqueous processing has been the major methodology for separations in the nuclear industry and thus far in nuclear waste treatment. Solvent extraction methods have the advantage of allowing a continuous separation, using counterflow modified PUREX systems with added specific separations at the backend, and are still the major technologies in use or planned for use. New organic and inorganic resins and new extractants are under development and may provide much improved separations with smaller amounts of secondary waste streams. Less developed at present are separation systems using membrane processes, natural agents, and magnetic fields.

Preparation of this chapter was supported by a contract from the U.S. DOE-BES Division of Chemical Science.

References

1. Hill, O. F., and Cooper, V. R. (1958) *Ind. Eng. Chem.* **50**, 599.
2. Bruce, F. R., Fletcher, J. M., and Hyman H. M. (eds) (1958) *Progress in Nuclear Energy* Series III, Process Chemistry, vol. 2, Pergamon Press, New York.
3. Flanary, J. R. (1954) *Reactor Sci. Technol.* **4**, 9.
4. National Research Council (1995) *Nuclear Wastes, Technologies for Separations and Transmutations*, National Academy Press, Washington, D. C., 151-152.
5. Rawlins, J. A., and Bager, H. R. (1990) *CURE: Clean Use of Reactor Energy*, Report WHC-EP-0268, Westinghouse.

6. Bauer, R. (1992) Cesium Cut from Radioactive Waste, *C&E News* **26**.

7. Horwitz, E. P., Dietz, M. L., and Fisher, D. E. (1992) *Solv. Extr. Ion Exch.* **8**, 557.

8. OECD (1997) *Actinide Separation Chemistry in Nuclear Waste Streams and Materials*, OECD Report NEA/NSC/DOC (97), 19.

9. Nash, K. L. (1993) Actinide Phosphonate Complexes in Aqueous Solutions, *Abstracts of Actinides-93 International Conference*, Santa Fe, New Mexico.

10. Weaver, B., and Kappelmann, F. A. (1964) *A New Method of Separating Americium and Curium from Lanthanides by Extraction from an Aqueous Solution of Aminopolyacetic Acid Complex with a Monoacidic Phosphate or Phosphonate*, Report ORNL-3559, Oak Ridge National Laboratory.

11. Persson, G. E., Syatesson, S., Wingefors, S., and Liljenzin, J. O. (1984) *Solv. Extr. Ion Exch.* **2**, 89.

12. Lloyd, M. H. (1963) *Nucl. Sci. Eng.* **17**, 452.

13. Musikas, C., and Hubert, H. (1987) *Solv. Etr. Ion. Exch.* **5**, 877.

14. Rosen, A. M., and Nikolotova, Z. I.(1991) *Radiokhimiya* **33**, 1.

15. Madic, C., (1994) Actinide Partitioning from High-Level Liquid Waste Using the Diamex Process, In *Proc. 4th Intern. Conf. on Nuclear Fuel Reprocessing and Waste Management*, RECOD '94, vol. 3.

16. Horwitz, E. P., and Schulz, W. W. (1999) Solvent Extraction in the Treatment of Acidic High-Level Liquid Waste: Where Do We Stand? In *Metal-Ion Separation and Preconcentration: Progress and Opportunities*. Bond, A. H., Dietz, M. L., and Rogers, R. D. (eds) ACS Symposium Series 716, Washington, D.C.

17. Musikas, C. (1985) Actinide/Lanthanide Group Separation Using Sulphur and Nitrogen Donor Extractants. In Choppin, G. R., Navratil, J. D., and Schulz W. W. (eds) *Actinide/Lanthanide Separations*, World Sci. Press, Philadelphia, 9.

18. Ensor, D. D., Jarvinen, G. D., and Smith, B. F. (1988) *Solv. Extr. Ion Exch.* **6**, 439.

19. Rais, J., Tachimori, S., and Selucky, P. (1964) Synergetic Extraction in Systems with Dicarbollide and Bidentate Phosphonate, *Sep. Sci. Technol.* **29**, 261.

20. Esimantovskii, V. M., Galkin, L. M., Lazarev, R. I., Lyubtsev, V.N., Romanovsky, V. N., and Shishkin, D. N. (1992) Technological tests of HAW Partitioning with the Use of Chlorinated Cobalt Dicarbollide (CHCODIC): Management of Secondary Wastes. In *Proc. Symp. on Waste Management*, Tucson, Arizona, 801.

21. Romanovsky, V. N. (1999) Review of Historical Development and Application of Separation Technologies in Russia. In *Chemical Separation Technologies and Related Methods of Nuclear Waste Management*, Choppin, G. R., and Khankhasayev, M. Kh. (eds) Kluwer Academic Publishers, Dordrecht.

22. Myasoedov, B. F., Karalova, Z. K., Nekrasova, V. V., and Rodinova, L. M (1980) *J. Inorg. Nucl. Chem* **42**, 1495.

23. Mahajan, G. R., Ray, M., Karekar, C. V., Rao, V. K., and Natarajan, P. R. (1985) Extraction of Plutonium from Alkaline Solutions by Quarternary Amine Aliquat-336. In *Radiochemistry and Radiation Chemistry Symposium*, Kanpur, India.

24. Hooper, E. W., Phillips, B. A., Dagnall, S. P., and Monkton, M. P. (1984) *An Assessment of the Application of Inorganic Ion Exchangers to the Treatment of Intermediate Level Wastes*, AERE-R 11088, AERE, Harwell Laboratory, England.

25. Chiarizia, R., Dorey, K. A., Horwitz, E. P., Alexandratos, S. D., and Jarvinen, G. D. (1996) *Solv. Extr. Ion. Exch.* **14**, 519.

26. Trochimczuk, A. W. (1999) Technology Needs for Actinide and Technetium Separations Based on Solvent Extraction, Ion Exchange, and Other Processes. In *Chemical Separation Technologies and Related Methods of Nuclear Waste Management*, Choppin, G. R., and Khankhasayev, M. Kh. (eds), Kluwer Academic Publishers, Dordrecht.

27. John, J., Sebesta, F., and Motl, A. (1999) Application of New Inorganic-Organic Composite Absorbers with Polyacrylonitrile Binding Matrix for Separation of Radionuclides from Liquid Radioactive Wastes. In *Chemical Separation Technologies and Related Methods of Nuclear*

Waste Management, Choppin, G. R., and Khankhasayev, M. Kh. (eds), Kluwer Academic Publishers, Dordrecht.

28. Nakajima, A., and Sakaguchi, T. (1990) *J. Chem. Tech. Biotechnol.* **47**, 31.

29. Sakaguchi, T., and Nakajima, A. (1987) *Sep. Sci. Tech.* **22**, 1609.

30. Whisenhunt, D. W., Neu, M. P., Xeu, J., Houk, Z., Hoffman, D. C., and Raymond, K. N. (1993) Thermodynamic Formation Constants for Actinide(IV) Ions with Sidereophores and Sidereophore Analogs, Abstract P127, *Actinides 93 Intern. Conf.*, Santa Fe, New Mexico.

31. National Research Council (1995) *Nuclear Wastes, Technologies for Separations and Transmutation*, National Academy Press, Washington, D.C.

32. Avens, L. R., Gallegos, U. F., and McFarlan, J. T. (1990) *Sepn. Sci. Tech* **25**, 1967.

33. Hoegler, J. M., and Bradshaw, W. M. (1989) *Magnetic Separation of DOE Wastes*, ORNL/TM 11117, Oak Ridge Nat. Lab.

Gregory R. Choppin is R. O. Lawton Distinguished Professor of Chemistry at Florida State University (FSU), Tallahassee, Florida, U.S.A. From 1968 to 1977, he was Chair of the Chemistry Department at FSU. He was a member of the U.S. National Research Council—National Academy of Sciences Board of Chemical Science and Technology, and is a member of the Board of Radioactive Waste Management, the Committee on Electrometallurgical Processing of DOE Spent Nuclear Fuel, the Committee on Remediation of Buried and Tank Wastes, and the Committee on a Long-Term Environmental Quality R&D Program in the U.S. DOE. He received his B.S. degree from Loyola University (New Orleans, LA), a Ph.D. in chemistry from University of Texas (Austin, TX), and Honorary D.Sc. degrees from Loyola University and from Chalmers University of Technology (Sweden). He has received awards from the American Institute of Chemistry, the American Chemical Society, the American Nuclear Society, and the British Royal Society of Chemistry.

CHAPTER II

THE ENVIRONMENTAL LEGACY OF THE COLD WAR: SITE CLEANUP IN THE UNITED STATES

TERESA FRYBERGER
U.S. Department of Energy
Washington, D.C. 20585, U.S.A.

ABSTRACT

Fifty years of nuclear technology and weapons development have produced an enormous environmental legacy in the United States. Nuclear technology contributed greatly to U.S. national security during the Cold War. The treatment of chemical and radioactive wastes and their impacts on the environment were of secondary concern to the production of nuclear weapons. Weapons production has left the U.S. with contaminated soil, surface water, and ground water, as well as large volumes of radioactive and chemical wastes, many of which are unique. No other industry or government activity has ever created wastes or environmental contamination of this nature [1,2]. Separation science and technology are critical to the weapons complex cleanup. Although progress has been made in both the science and technology regarding the cleanup over the last decade, a great deal remains to be done. Separation needs and challenges for two DOE waste types, high-level tank wastes at Hanford and dilute aqueous waste, will be discussed in this chapter.

1. Introduction

The U.S. Department of Energy (DOE) is the government agency responsible for managing the nuclear weapons complex, which includes 120 million square foot of buildings and facilities and 2.3 million acres of land used for testing, research, and production of nuclear weapons. The DOE's cleanup challenge encompasses 3,700 contaminated sites in 34 states; more than 1,000 million gallons of radioactive and mixed wastes stored in 322 tanks; 3 million cubic meters of radioactive or hazardous buried wastes; 250 million cubic meters of contaminated soils; more than 600 billion gallons of contaminated ground water; and over 2,000 facilities requiring decontamination and decommissioning. DOE currently estimates that the cleanup costs will range from $200 to $350 billion over 75 years [1,2,3].

Essentially all types of DOE defense wastes require separation methods to concentrate the contaminants and to purify waste streams before they can be released to the environment or downgraded to a less costly form for disposal. Although separation technologies for nuclear fuel reprocessing have been in use for many decades, the unique chemistry and complexity of DOE radioactive wastes preclude the direct transfer of reprocessing technologies. Existing reprocessing technologies can be adapted for treatment of some DOE wastes, but entirely new approaches are needed in most cases. Because of the need to protect the environment, separation technologies that minimize secondary wastes and use environmentally benign chemicals are required.

2. Nuclear Weapons Production Waste Streams

Over the last 50 years, development of nuclear weapons by the United States has resulted in a vast research, production, and testing network known as the nuclear weapons complex. This complex is currently managed by the U.S. Department of Energy. As shown in Figure 2-1, the U.S. nuclear weapons complex comprises dozens of industrial and research facilities across the country. Figure 2-1 also illustrates the various steps in the production of nuclear weapons.

Like most industrial and manufacturing operations, the nuclear weapons complex has generated waste, pollution, and contamination. Many of the environmental problems created by weapons production, however, are unlike those associated with any other industry. They include radiation hazards, unprecedented volumes of wastes, contaminated water and soils, and a large number of contaminated structures, such as reactors, chemical plants, and storage tanks. Table 2-1 provides the U.S. definitions of waste types generated

---

TABLE 2-1. DOE Environmental Legacy: Nuclear Waste Types

Spent fuel: fuel elements and irradiated targets (designed "reactor-irradiated nuclear material" and often called simply "spent fuel") from reactors. The department's spent fuel is not categorized as waste, but it is highly radioactive and must be stored in special facilities that shield and cool the material.

High-level waste (HLW): material generated by the reprocessing of spent fuel and irradiated targets. Most of the department's high-level waste came from the production of plutonium. A smaller fraction is related to the recovery of enriched uranium from naval reactor fuel. This waste typically contains highly radioactive, short-lived fission products as well as long-lived isotopes, hazardous chemicals, and toxic heavy metals. It must be isolated from the environment for thousands of years. Liquid high-level waste is typically stored in large tanks, while waste in powered form is stored in bins.

Transuranic waste (TWU): waste generated during nuclear weapons production, fuel reprocessing, and other activities involving long-lived, transuranic elements. It contains plutonium, americium, and other elements with atomic numbers higher than that of uranium. Some of these isotopes have half-lives of tens of thousands of years, thus requiring very long-term isolation.

Low-level waste (LLW): any radioactive waste that does not fall into one of the other categories. Every process involving radioactive materials produces it. Low-level waste contains small amounts of radioactivity in large volumes of material. Some wastes in this category (e.g., irradiated metal parts from reactors) can be more radioactive per unit volume than the average high-level waste from nuclear weapons production. Most low-level waste has been buried near the earth's surface. A limited inventory remains stored in boxes and drums.

Mixed waste: waste that contains both radioactive and chemically hazardous materials. All high-level and transuranic wastes are managed as a mixed waste. Some low-level waste is mixed waste.

Uranium-mill tailings: large volumes of material left from uranium mining and milling. While this material is not categorized as waste, tailings are of concern both because they emit radon and because they are usually contaminated with toxic heavy metals, including lead, vanadium, and molybdenum.

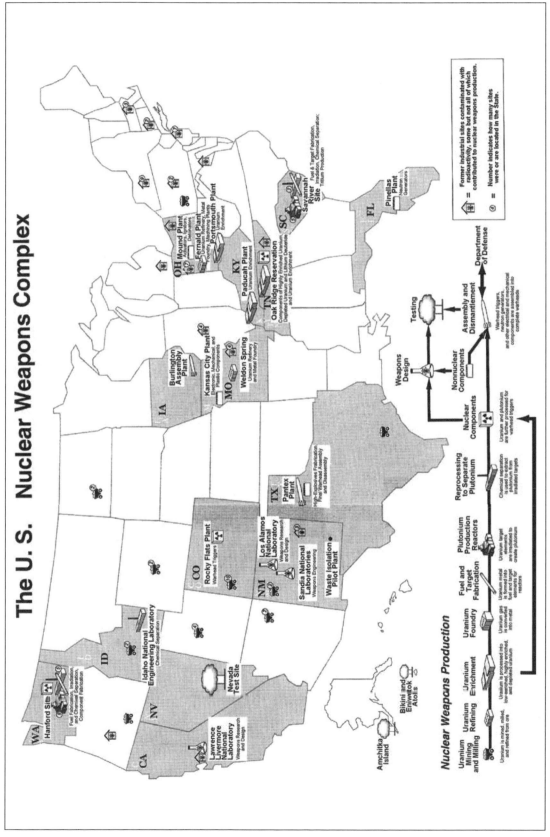

Figure 2-1. The U.S. nuclear weapons complex; and, at bottom, sequence of steps in production of weapons [1].

from the production of weapons, and Figure 2-2 shows the relative volumes of wastes generated by each of the weapons production steps. The impacts to the environment are illustrated in Figure 2-3, which depicts the expected contaminant sources and pathways at a major weapons production site. References 1 and 2 provide excellent overviews of the history of the complex and the waste legacy it created.

In the late 1980s, with the end of the Cold War and the increasing awareness of its environmental impacts, the DOE began a massive weapons complex cleanup. Estimates for its cost are in the hundreds of billions of dollars over the next 75 years [3]. An environmental "project" of such scope is unprecedented anywhere in the world. Because no industry or government has ever tackled most of these environmental issues, a serious lack of both knowledge and technology existed from the outset. For example, little was known about how contaminants (radioactive or otherwise) move in the environment, about how these contaminants impact human health or ecosystems at environmental levels, about how the chemistry of wastes affects treatment and waste form performance, and how waste forms and repositories perform over thousands of years. All of these factors profoundly impact how cleanup activities are or should be prioritized and pose tremendous challenges for cleanup technology. Though progress has been made in both the science and technology for the cleanup over the last decade, a great deal remains to be done. Lessons from the DOE cleanup should help shape a better future everywhere for environmental remediation and waste management practices.

In the following sections, the overall benefits and challenges for separations technologies in the weapons complex cleanup are described. This will be followed by a more specific discussion of the characteristics of and separations challenges posed by high-level tank wastes stored at the Hanford site and dilute aqueous wastes found across the DOE complex. Both existing and emerging separation techniques are discussed in Chapters I and IV.

3. Separation Needs and Requirements

Removal of radionuclides and other contaminants from the less toxic portions of DOE wastes can:

- Reduce waste volumes;
- Improve characteristics of final waste forms for disposal (e.g., transform long-lived radioisotopes to shorter-lived ones);
- Ease waste handling requirements;
- Recycle process reagents and reduce generation of new wastes during cleanup operations; and
- Reduce long-term health risks.

All of these benefits have the potential for impressive cost savings. For example, at Hanford alone an estimated tens of billions of dollars could be saved by using separations to further concentrate radionuclides, thereby reducing the volume of tank waste that must be disposed of in a geological repository. But, improved separations technologies *must* be developed before we can even treat certain wastes. At the Savannah River Site, HLW was to be treated using an inadequately developed and tested separations scheme. Ten years and billions of dollars later, the DOE is now forced to revisit their entire HLW pretreatment plan at that site [4].

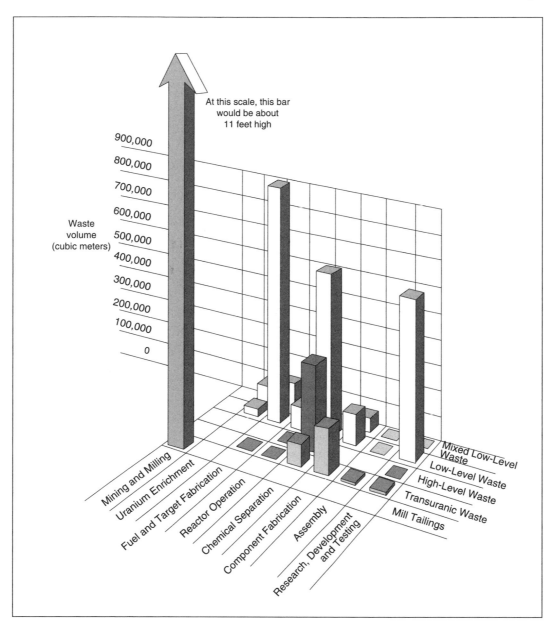

Figure 2-2. Volume of wastes and other byproducts generated by nuclear weapons production activities during the Cold War [1].

Although separation science provides potential benefits, proven technologies do not exist for many DOE waste streams, and certainly not for the most challenging wastes such as HLW. The nuclear technical community has a great deal of familiarity with radio-chemical separations from years of spent fuel reprocessing, but the technical challenges for cleanup separations are very different from those of reprocessing. First, the chemical media are very different. The separations steps in reprocessing (which ironically produced much of DOE's most challenging waste stream) are performed on a well-characterized acidic processing stream. The purpose of reprocessing is to *recover* plutonium or uranium

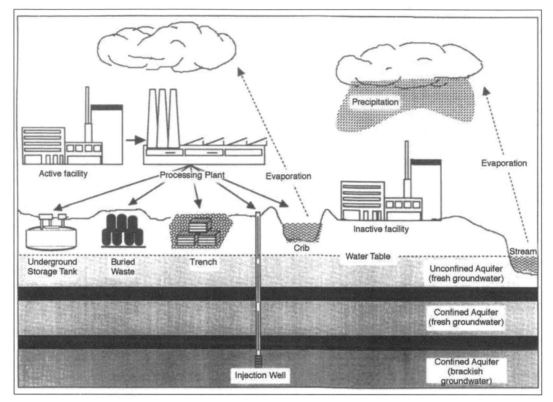

Figure 2-3. Examples of contamination sources and pathways at a weapons production facility [2].

for reuse, not to reach the low levels necessary to decontaminate waste. The waste media requiring separations in the cleanup effort are far more varied and include:

- Dilute aqueous streams;
- Acidic, alkaline, and neutral waste streams;
- High ionic strength wastes;
- Chemically complex or poorly characterized wastes;
- Solid precipitates and residues;
- Soils and buried wastes; and
- Wastes requiring the removal of several species.

Examples of species that require removal from wastes, or soils and ground water include:

- Transuranic elements (e.g., Pu, Np, Am);
- Strong radiation sources (^{137}Cs);
- High heat generators (^{137}Cs, ^{90}Sr);
- Long-lived soluble fission products (e.g., ^{99}Tc);
- Hazardous metals (e.g., Hg, Ni, Pb); and

- Chemicals or metals that complicate production of final waste forms, such as borosilicate glass (e.g., organics, noble metals).

These characteristics pose serious challenges for separation agents and technologies. Today's separation technologies must meet more stringent environmental standards than those in the past (i.e., they must minimize secondary wastes and use chemicals that are environmentally "benign").

4. High-Level Tank Wastes at the Hanford Site

The Hanford Site in the state of Washington is home to 177 underground storage tanks that store 208 million liters of high-level waste containing about 200 megacuries of radioactivity (Figures 2-4 and 2-5). An excellent overview of the Hanford site and its HLW is provided in reference 5 (see also references 6 and 7). Most of this HLW originated from the reprocessing of spent nuclear fuel to recover plutonium. During reprocessing, spent fuel is dissolved in acid, and plutonium is separated from the acidic solution—in earlier days by precipitation, later by solvent extraction. The remaining acidic raffinate becomes the HLW, which includes most of the fission products and is highly radioactive. To save costs, and because stainless steel was not readily available during World War II, this waste was neutralized by adding an excess of sodium hydroxide to create a highly alkaline (pH range from 9 to 14) solution. Sodium nitrite was also added as a corrosion inhibitor. The waste was then stored in large carbon steel, underground storage tanks, ranging in size from 55,000 gallons to 1.1 million gallons. Over the years, various operations took place in different tanks, such as addition of chemicals (e.g., organic and ferrocyanide species) and transfer of wastes to different tanks. The HLW waste phases and their approximate chemical content are illustrated in Figure 2-5. Note that HLW was treated similarly at the Savannah River Site, and the issues there are not unlike those at

- **Single Shell Tanks**
  - 149 tanks
  - 35M gallons/190K tons of wastes
  - 132M curies of radioactivity (75% Sr-90, 24% Cs-137)
  - 65 "leakers"
- **Double Shell Tanks**
  - 28 tanks
  - 20M gallons/50K tons of waste 82M curies of radioactivity (72% Cs-137, 27% Sr-90)

Figure 2-4. High-level waste tanks under construction at the Hanford site.

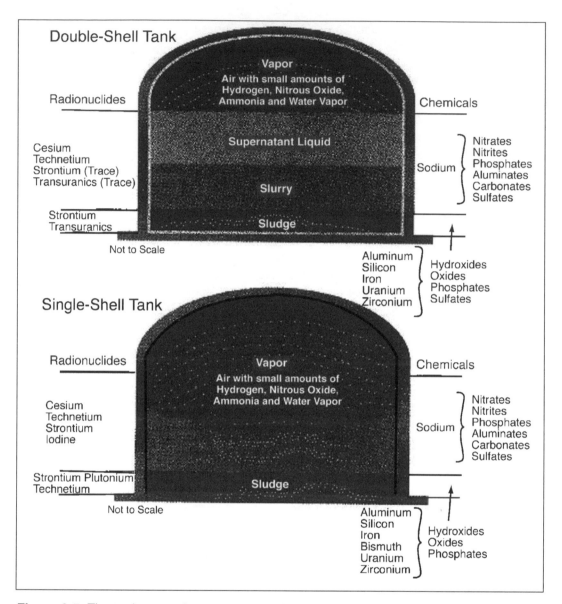

Figure 2-5. The tanks contain numerous radionuclides and chemicals that have separated into blended layers. The contents of any individual tank can vary significantly from these two idealized illustrations [5].

Hanford, although slightly different waste treatment approaches are being taken at the two sites.

As one can see from Figures 2-4 and 2-5, the chemistry of the tank wastes is highly complex. To add to this challenge, the waste has been stored much longer than originally planned. Many of the tanks are leaking, and the waste contents are variable and not completely characterized.

The basic (simplified) flowsheet for Hanford HLW treatment and disposal is shown in Figure 2-6 (this is very similar to the SRS plan). From that figure, one can see that removal

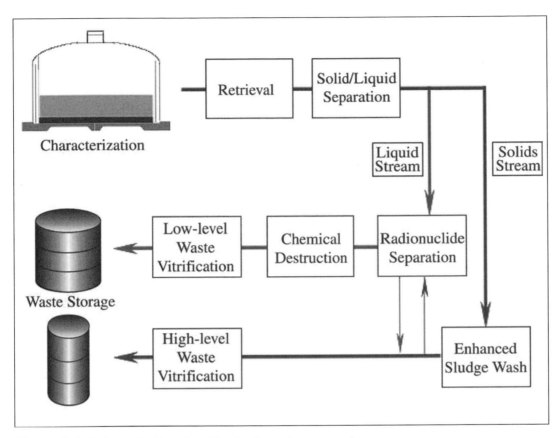

Figure 2-6. Schematic flowsheet for tank waste processing.

of radionuclides from the liquid portions of the waste (supernate and dissolved saltcake) is planned. This allows the liquid portions of the waste to qualify for disposal as LLW, thereby decreasing the volume of costly HLW. The primary constituents for removal are cesium and strontium, but some transuranic species must also be removed. Some of the separations challenges that Hanford (and Savannah River) face include:

- The radioactive species (Cs, Sr) to be removed are exceedingly dilute, requiring separations agents of very high affinity.

- Cesium must be separated from chemically similar sodium, which is present at orders of magnitude higher concentration than cesium, requiring highly selective sequestrants.

- The waste is highly alkaline, requiring separation agents and media that are durable in caustic solution.

- Because cesium is highly radioactive and a high heat generator (as is strontium), the chemicals and materials used to concentrate them must be highly durable to both radiation and heat.

- The complexity and huge volumes of waste to be treated may necessitate long residence times of the concentrated radionuclides in their separations media (i.e., weeks to months), requiring even greater durability, unless new storage tanks are built. It may be possible to avoid such long residence times by careful process design.

- The chemical variability and poor characterization of the wastes will require careful characterization as the waste goes into separations treatment, and this must be coupled with a highly flexible separations scheme that can accommodate the changing chemical makeup of the wastes.

A number of ion-exchange approaches have been considered for cesium removal from the Hanford tank wastes using existing materials technology, such as CS-100 (an organic resin) to advanced materials such as crystalline silicotitanate and new "designer " ligands [8]. To date, none of these approaches have been demonstrated at pilot scale on actual Hanford tank waste. For strontium and transuranic removal, the baseline at both Hanford and Savannah River is precipitation using monosodium titanate. Although this technology has been used at Savannah River at full scale on actual waste, some technical difficulties with rates of radionuclide removal and solid/liquid separations remain to be resolved [4].

5. Dilute Aqueous Wastes across the DOE Complex

Dilute aqueous wastes come from a wider variety of sources and have far greater volumes than HLWs:

- Virtually every step in weapons production generated low-level waste, most of which is protective clothing, rags, contaminated equipment, etc.; but a significant portion is aqueous;

- 5,700 plumes contaminating soil and ground water have been identified (though not all will require or use separation treatment);

- Contaminated spent fuel storage basins exist at Hanford and other sites;

- Newly generated aqueous wastes will result from cleanup activities; these are secondary wastes generated from decontamination of facilities, from waste processing, from fuel handling, and from pumping and treatment operations.

Separation of contaminants from dilute aqueous wastes provides the same major benefit as separation of contaminants from HLW, i.e., reduction in waste volumes. Some requirements for separations are also shared by both waste types, such as the need for separation agents with high affinity, because of the very low concentrations of contaminants in both media. Unlike HLW, however, dilute aqueous wastes are rarely as chemically complex or as harsh, so that the durability requirements are not so stringent. Many of the dilute aqueous waste streams are in the environment—in soils or ground water—creating challenges for delivery of separations agents to the contaminants and recovery of the contaminants. Perhaps the greatest challenge for separations from dilute aqueous streams is their tremendous volume. Very high throughput and fast kinetics are needed. Separation processes must be economical, and sequestrants must be reusable with small losses.

The utmost consideration when investing in technologies for cleanup must be human health. Despite the fact that dilute aqueous wastes are less toxic than HLW, they are not of lesser importance. They are frequently more accessible to humans and ecosystems through environmental pathways such as waterways. What we learn from developing separations technologies for these waste streams will be widely applicable to a broad range of environmental problems, from mining and industrial pollution, to manufacturing processes used in the chemical and other industries, to waste management approaches for the future.

6. Conclusions

New separation technologies are needed to meet DOE cleanup goals. In some cases, improvements or modifications to existing technologies will suffice. But in many cases, new sequestrants, new delivery systems, and innovative process designs are needed. Over the last decade, a great deal of progress has occurred in the development of new separations agents and concepts [8], but few of these efforts have been taken beyond the laboratory stage to be tested at pilot scale on real wastes. For success to occur, we must continue to invest in alternatives beyond the laboratory level and until baseline technologies are *proven* in actual operation. Otherwise a failure in a technology can result in the loss of a great deal of time and money [3].

Many arguments and much discussion have occurred in the DOE community about the extent of separations that is needed or desired for various wastes. For many waste streams, such as HLW, DOE has opted for a minimal separations approach. By more extensive separation of Hanford HLW (e.g., by dissolving and treating the sludges), however, we could potentially further concentrate radionuclides, thereby reducing volume (and hence cost) of the HLW portion. This would require extensive chemical processing and technology development, potentially driving up total costs and extending schedules.

Thus the proper extent of separations for a given waste is a complex issue and requires careful long-term analysis of *the entire treatment and disposal scheme*. Some of the questions that must be addressed in evaluating separation technologies for a given waste are:

- Will additional hazards be created?
- What are the costs and schedule impacts to the overall plan?
- How will separation processes impact other treatment steps?
- What are desired end-states, so that performance requirements can be defined?
- Is there a large enough market to attract industrial production of sequestrants and materials at reasonable cost?
- Can advanced separations technologies, proven at laboratory scale, be successful in the field at full scale and on actual waste?

The DOE weapons complex cleanup is a difficult endeavor, and contaminant separations are only one of many technical challenges. But despite the difficulties and false starts, it is a tremendous opportunity to pave the way for understanding how to deal with (or prevent) the many other environmental threats that we face now and will face in the future.

References

1. *Closing of the Circle on the Splitting of the Atom: The Environmental Legacy of Nuclear Weapons Production in the United States and What the Department of Energy is Doing about It*. U.S. Department of Energy, Office of Environmental Management, Washington, D.C. (1996).

2. *Linking Legacies: Connecting the Cold War Nuclear Weapons Production Processes to Their Environmental Consequences.* U.S. Department of Energy, Office of Environmental Management, Washington, D.C. (1997).
3. *From Cleanup to Stewardship.* U.S. Department of Energy, Office of Environmental Management, Washington, D.C. (1999).
4. *Alternatives to High-Level Waste Salt Processing at the Savannah River Site.* National Academy Press, Washington, D.C. (2000).
5. Gephart, R.E., and Lundgren, R.E. (1998) *Hanford Tank Cleanup: A Guide to Understanding the Technical Issues.* Battelle Press, Columbus, Ohio.
6. *The Hanford Tanks: Environmental Impacts and Policy Choices.* National Academy Press, Washington, D.C. (1996).
7. *An End-State Methodology for Identifying Technology Needs for Environmental Management, with an Example from the Hanford Site Tanks.* National Academy Press, Washington, D.C. (1999).
8. *Efficient Separations and Processing Program: Technology Summary.* Department of Energy, Office of Environmental Management, Washington, D.C. (1995).

Teresa Fryberger is the Associate Deputy Assistant Secretary for Science and Technology within the Department of Energy's Environmental Management program. In this capacity, Dr. Fryberger manages a national research and development program that furnishes innovative environmental cleanup technologies for use in remediating the department's weapons complex.

Prior to this position, Dr. Fryberger was the Associate Laboratory Director for the Energy, Environment, and National Security Directorate at Brookhaven National Laboratory (BNL). As head of three science departments at BNL, she managed and developed a diverse program in environmental sciences, energy sciences, national security, as well as applied chemistry and materials science.

Dr. Fryberger has also been Senior Deputy Director of Pacific Northwest National Laboratory's William R. Wiley Environmental Molecular Sciences Laboratory, where she was responsible for managing environmental science programs and providing strategic direction for the overall management of this then-new National Scientific User Facility.

Earlier in her career, Dr. Fryberger managed national scientific programs at the Department of Energy, was an associate editor at *Science*, and was a research chemist and National Research Council Postdoctoral Fellow at the National Institute for Science and Technology. Dr. Fryberger has organized and chaired over fifty national and international meetings of professional societies, of the Department of Energy, and of technical organizations. She has also served on numerous advisory and review committees for national laboratories, the Department of Energy, and universities. She earned her Ph.D. in Physical Chemistry from Northwestern University and her B.S. in Chemistry from the University of Oklahoma.

CHAPTER **III**

ENVIRONMENTAL IMPACTS OF SEPARATION TECHNOLOGIES IN RUSSIA

BORIS F. MYASOEDOV
V.I. Vernadsky Institute of Geochemistry and Analytical Chemistry
Russian Academy of Sciences
117975 Moscow, Russia

ABSTRACT

The sources of radionuclides that enter the environment are examined including long-lived transuranic elements from different stages of the nuclear fuel cycle. Radioactive substance contamination from natural and man-made sources is analyzed for different regions of Russia. The potential danger of long-lived transuranic radionuclides in the wastes of nuclear fuel cycle plants is also pointed out. Data related to the series of nuclear weapons tests near Semipalatinsk and Novaya Zemlya are presented.

An overview is given of the current radioecological situation around the reprocessing plant at Mayak, which was constructed more than 50 years ago for the production of plutonium for military purposes. The following topics are considered: the highly contaminated Lake Karachay, contaminated artificial water reservoirs, and solid radioactive wastes and their vitrification.

Russian R&D on recovery of radionuclides from radioactive wastes is also discussed. New approaches, methods, and tools developed at the Vernadsky Institute for the identification of radionuclides in samples from the impact zone of the Mayak plant are described, and data on distribution, occurrence forms, and migration processes of ^{90}Sr, ^{137}Cs, ^{237}Np, ^{239}Pu, and ^{241}Am in aquatic and terrestrial ecosystems are presented.

1. Introduction

Several periods can be distinguished in the history of radiochemistry [1]. The first period, when radiochemistry became an independent scientific endeavor, was closely related to the discovery of new radioactive elements and an understanding of the principal laws of radioactive substance behavior.

During the second period (1940s to 1960s), radiochemical investigations were focused on the practical utilization of nuclear energy, the study of the chemical properties of artificially produced elements, the development of technology for processing the irradiated nuclear fuel, and the resolution of the problem of radioactive wastes disposal. In this period, the PUREX process was elaborated for separation of plutonium from irradiated uranium fuel. Unfortunately both in the former USSR and in the U.S.A., no attention was paid during this period to problems of radioactive waste management.

After the 1960s, the third modern period of radiochemistry development started. During this third period, society began to realize the global character of the consequences of this activity. The main attention shifted to the problems of continued development, including such aspects as remediation of polluted territories, study of the behavior of radionuclides in nature, reduction of the amount of unavoidable (for the current nuclear technology) radioactive wastes, development of the technology for long-term radioactive waste storage, and many related issues.

Environmental problems are among the most important issues facing society. Solutions to these problems are aggravated by heavy nuclear pollution in many regions. Artificial radionuclides have entered the environment from atmospheric and underground tests of nuclear weapons, from activities of nuclear power plants with nuclear fuel cycles producing and accumulating weapon-grade plutonium (in the U.S.A.: Hanford, Savannah River and Idaho facilities; in the former USSR: Mayak Production Association in Ozersk, Siberian Chemical Plant in Seversk, Mining and Chemical Plant in Krasnoyarsk-26), from accidents occurring at nuclear power plants, and from the unsanctioned submersion and disposal of nuclear waste from ships and submarines into the oceans. Rehabilitation of territories contaminated with radionuclides is now recognized as an important environmental, economic, and social problem.

2. Natural Background Radiation

All organisms on Earth, including humans, are exposed to radiation from space and from natural radionuclides that occur in the atmosphere, water, soil, rocks, and some types of food, as well as construction and other materials. Radon, which forms through the decay of natural uranium, exerts the most influence on living organisms. Radon exposure is responsible for 55% of the total annual dose of background radiation received by humans; internal irradiation by natural radionuclides (^{40}K, ^{232}Th, ^{235}U, and ^{238}U) contributes 11%; space radiation and cosmogenic radionuclides (^{3}H, ^{14}C) contributes 8%; and radioactivity from rocks and minerals contributes 8%. The remainder of the effective equivalent dose (Figure 3-1) is received through the use of active isotopes in various types of medical diagnostics (15%) and from the activity of artificial radionuclides (3%) [2].

3. The Main Sources of Nuclear Pollution in Some Regions of Russia

The influx of artificial radionuclides into the atmosphere began with the first nuclear explosions in 1945 and grew with the ensuing development of the nuclear industry. Table 3-1 shows estimates of radionuclide emission from various sources (in pBq) and radiation doses obtained by the population from these sources (in man-Sv).

3.1. Nuclear tests

Nuclear weapons testing has resulted in the discharge into the environment of a considerable amount of uranium, plutonium, and their fission products, which were distributed as aerosols and gases across large distances and which have produced most of

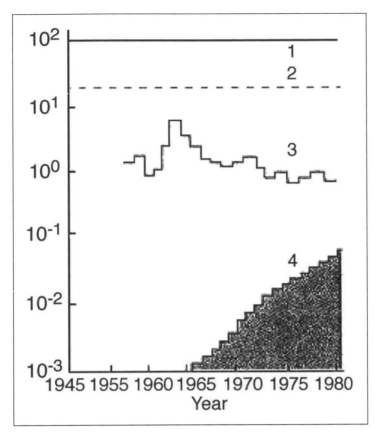

Figure 3-1.
Annual effective equivalent dose (in %) due to medical irradiation (2), nuclear explosions in the atmosphere (3), nuclear energy production and use of radioactive materials (4), compared with the mean annual effective dose of radiation from natural sources (1) [2].

the anthropogenic radioactive background on the Earth's surface, especially in the Northern Hemisphere.

Nuclear tests in the atmosphere are responsible for the vast majority (95%) of artificial radionuclides in the environment. The figure was highest in 1963, when it accounted for ~7% of the annual irradiation from natural sources; this fraction is currently less than 1% (Figure 3-1). By 1980, 455 tests of nuclear weapons had been conducted worldwide in the atmosphere, with a total capacity of 545 Megatons (Mt), including 217 Mt from fission reactions and 238 Mt from fusion reactions (Table 3-2). Before 1972 and the signing of the Moscow Declaration on the prohibition of nuclear tests in the atmosphere, on the ground, and in the ocean, the nuclear countries carried out nuclear test explosions in the atmosphere with the following capacities (in Mt): U.S.A., 96.72; USSR, 85.46; Great Britain, 10.50; France, 10.87; and China, 12.82 [4].

Radioactive products formed during nuclear explosions either precipitate near the explosion site (~12%), or are retained in the troposphere and later fall to the ground (~10%), or enter the stratosphere and fall all over the Earth (78%). The main portion, however, typically lands in the same hemisphere as the explosion.

A nuclear explosion of a capacity of 1 Mt produces $3.7'10^{15}$ Bq ^{90}Sr and $6.2'10^{15}$ Bq ^{137}Cs so that during the whole period of nuclear tests, $~0.5'10^{18}$ Bq ^{90}Sr and $0.8'10^{18}$ Bq ^{137}Cs as well as 5-10 tons of the environmentally most hazardous element, plutonium, entered the atmosphere.

In the former USSR, nuclear tests in the atmosphere were carried out from 1949 to 1962 in the Semipalatinsk area (124 explosions with a total capacity of 6.3 Mt releasing 0.1MCi ^{90}Sr, and 0.2MCi ^{137}Cs), and on the island Novaya Zemlya (87 explosions,

TABLE 3-1. Estimates of the Emission of Artificial Radionuclides and Cumulative Effective Dose [3]

Source	Emission, PBq						Total effective dose, man-Sv	
	^{3}H	^{14}C	noble gases	^{90}Sr	^{131}I	^{137}Cs	local and regional	global
Nuclear tests in the atmosphere								
Global	24×10^4	220		604	65×10^4	910		223×10^5
Local								
Semipalatinsk							4600	
Nevada							500[b]	
Australia							700	
Pacific Ocean							160[b]	
Underground nuclear explosions			50		15		200	
Production of nuclear weapons								
Initial tests								
Hanford							8000[a]	
Chelyabinsk							15000[e]	
Later tests							1000	10^4
Nuclear energy production								
Mining technology							2700	
Reactor operation	140	1.1	3200		0.4		3700	
Fuel reprocessing	57	0.3	1200	6.9	0.004	40	4600	
Fuel cycle							3×10^{5a}	10^5
Production and use of radioisotopes	2.6	1.0	52		6.0		2000	8×10^4
Accidents								
Three Mile Island		370		0.0006		40		
Chernobyl					630	70		60×10^4
Kyshtym				5.4		0.04	2500	
Windscale			1.2		0.7	0.02	2000	
Palomares							3	
Tule							0	
SNAP 9A								2100
Kosmos-954				0.003	0.2	0.003		20
Tundad Juarez							150	
Mohammedia							80	
Goiania						0.05	60	
Total							38×10^4	231×10^5
Cumulative effective dose, man-Sv							235×10^5	

[a] Normalized to 10000 years; [b] external dose only; [c] after discharge of ^{131}I into the atmosphere; [d] after outflow of radionuclides to the Techa River; [e] total dose due to extraction of ^{222}Rn from waste.

TABLE 3-2. Nuclear Explosions in the Atmosphere [3]

Period	Number of explosions	Capacity, Mt		Period	Number of explosions	Capacity, Mt	
		fission	total			fission	total
1945–1951	26	0.8	0.8	1964–1969	22	10.6	15.5
1952–1954	31	37	60	1970–1974	34	10	12.2
1955–1956	44	14	31	1975	0	0	0
1957–1958	128	40	81	1976–1980	7	2.9	4.8
1959–1960	3	0.1	0.1	1981–1990	0	Not performed*	
1961–1962	128	102	340	1945–1990	423	217.4	545.4
1963	0	0.0	0.0				

* In 1995, France performed new tests.

including 3 from a submarine, with a total capacity of 250 Mt releasing 4.2 MCi ^{90}Sr and 8.0 MCi of ^{137}Cs).

Eighty-one commercial nuclear explosions have also occurred in Russia, including some aimed at intensifying oil production, and others for seismic surveys and construction purposes. Some of the explosions, e.g., Kristall (1974) and Kraton-3 (1978) in Yakutia, resulted in the emission of radioactive vapor and gas into the atmosphere, including a small amount of plutonium.

The density of pollution (in Bq/m^2) with the most dangerous radionuclide, plutonium, from nuclear weapons tests and other sources, varies by country as follows [5]: Germany, 67-148; Ireland, 33-127; Great Britain, 33-122; Russia: Leningrad Nuclear Power Plant region, 14-262, Beloyarskaya Nuclear Power Plant region, 116-183; Mayak, up to 1400; Ukraine: Chernobyl Nuclear Power Plant (30 km zone), 3700; and Japan, 90.

3.2. Nuclear power plants

The commercial use of atomic energy in the former USSR was launched in 1954 with the installation of the 5-MW nuclear power station in Obninsk. According to the International Atomic Energy Agency, by the time of the Chernobyl accident, 417 nuclear power units in 26 countries were functioning and producing 285×10^9 W of electrical energy.

By the end of 1998, the 29 nuclear power units of Russia's nine nuclear power plants with an established electric capacity of 21 242 MW accounted for 11.8% of total Russia's electrical energy supply. Although the work of a nuclear power plant is environmentally less hazardous than that of a power station that uses coal, nuclear power plants do release a certain amount of long-lived radionuclides (^{85}Kr, ^{3}H, ^{14}C, and ^{129}I) into the atmosphere [6].

The main problems in the development of nuclear energy involve ensuring safe operation and waste management, especially of high-level wastes. In this context, potential use of the thorium fuel cycle with fluoride melts, utilization of highly enriched (up to 99.9%) uranium-235, new technological approaches to the reprocessing of spent nuclear fuel, and transmutations of nuclear waste are now being considered in several countries (see, e.g., Chapter VII).

3.3. Accidents at nuclear power plants

During the Chernobyl accident of 26 April 1986, ~3.5% of the total amount of radionuclides accumulated in the reactor of the fourth unit (~50 MCi) were released to the environment [7, 8]. The fallout of radioactive materials resulted in considerable radioactive pollution of the environment, mainly in the European part of the former USSR (Table 3-3). However, the inflow of radionuclides into the biosphere from the Chernobyl accident accounted for not more than 3% of the amount of radionuclides released to the environment by nuclear weapons tests (Table 3-1).

The Central Black Soil Regions were the most heavily contaminated by the accident. The western regions of the Bryansk Oblast received the greatest impact. Some of these regions show contamination of 40 Ci/km^2. The largest territories contaminated with ^{137}Cs are the Bryansk and Tula Oblasts where 45% of that territory has a level of ^{137}Cs higher than 1 Ci/km^2 (57,650 km^2, 1.6% of the area of the Russian Federation) [9]. Except for several localities in the Bryansk and Kaluga Oblasts, the exposure acquired by the population is mainly (^{3}90%) from natural background radiation and not from precipitated artificial isotopes.

Thus, according to the data in references 2 and 7, the Chernobyl accident did not significantly increase European Russia's total radiation exposure, and it may be assumed that the dose due to nuclear power will not exceed 1% of the natural background radiation by the year 2000 (Figure 3-1).

TABLE 3-3. Emission of Long-Lived Radionuclides due to the Chernobyl Accident and their Precipitation in the European Part of the USSR [9]

Radio-nuclide	Emission, PBq	Activity of radionuclides deposited in the European part of the USSR, PBq	Radio-nuclide	Emission, PBq	Activity of radionuclides deposited in the European part of the USSR, PBq
^{137}Cs	100	30	$^{239, 240}$Pu	0.055	0.05
^{134}Cs	50	15	^{238}Pu	0.025	0.02
^{106}Ru	35	25	^{241}Pu	5	4
^{144}Ce	90	75	^{241}Am	0.006	0.005
^{110m}Ag	1.5	0.5	^{242}Cm	0.6	0.55
^{125}Sb	3	2	$^{243, 244}$Cm	0.006	0.005

3.4. Plutonium production for military purposes and spent nuclear fuel reprocessing plants

Reprocessing of irradiated nuclear fuel contributes significantly to the radioactive pollution of the environment although it makes high-level wastes more manageable because it reduces their volume and provides more options for future management. Similar to nuclear weapon tests, reprocessing contaminates the biosphere with long-lived a-nuclides of transuranium elements.

Plutonium production in the USSR began in Ozersk (Chelyabinsk-40) at Plant 817, which was transformed later to the Mayak Industrial Association (Chelyabinsk-65). The first uranium-graphite reactor, Annushka, achieved full operating capacity on 19 June 1948. Water from Lake Kyzyl-Tash was used to cool the reactor. Numerous accidents resulted in the radioactive pollution of the lake. Flow-type cooling was also used in other industrial reactors that were installed later at Plant 817 [10].

The largest outflow of radioactive waste into the environment began with the start-up of the first radiochemical plant, known as Plant B (February 1949). Because of flaws in technology, even the wastewater from Plant B was poured directly into the Techa River, and an outflow of 1,000 Ci per day was allowed. The greatest amount of radioactive waste was released into the Techa River in 1950–1951, with a total of 2.75×10^6 Ci released between 1949 and 1956 [11]. The river flood plain was polluted over a distance of l00 km and ~125,000 persons, including 28,000 who received considerable doses, were irradiated. From 1958 on, the largest amount (more than 120 MCi) of medium-level nuclear waste (MLW) accumulated at Lake Karachai (Reservoir 9). In 1957, an explosion in one of the reservoirs of liquid high-level nuclear waste (HLW) resulted in the release of 2.0×10^6 Ci into the environment and the formation of the relatively narrow and long East Urals Radioactive Trail (EURT) (1000 km at a pollution density of 2 Ci ^{90}Sr per square kilometer). Three hundred thousand people were irradiated, and 10,000 of them received considerable doses [12].

In 1967, wind dispersed dust containing 600 Ci of radioactive compounds from the Karachai region, resulting in contamination of a 30-km^2 territory at a density of 2 Ci ^{90}Sr/km^2 and irradiation of 40,000 people. In order to prevent further dispersal of the radioactive dust, the area of the lake was diminished from 51 ha in 1962 to 15 ha in 1993 by filling it with gravel and rocks.

However, Lake Karachai is connected to ground water, which has become contaminated. During the exploitation period of the reservoir, the lake supplied 3.5×10^6 m^2 of industrial solutions to ground water with ~70,000 Ci ^{90}Sr, 20,000 Ci ^{137}Cs, 660,000 Ci ^{106}Ru, 100,000 Ci ^{3}H, and a considerable amount of uranium, neptunium, and plutonium. The dynamics of radionuclide dispersion in ground water is an urgent scientific problem [13].

In addition to the already described sources of radionuclide discharge to the environment, 600×10^6 Ci of liquid HLW accumulated during the 45-year period when Mayak was producing nuclear weapons. This waste could not be vitrified because of its complex chemical composition. About 13×10^6 Ci of solid HLW was stored in 24 reinforced concrete surface reservoirs and about 30,000 Ci of MLW and LLW in 200 ground reservoirs. About 400×10^6 m^3 of contaminated water was stored in industrial ponds on the Techa River (mainly ponds 10 and 11 with areas of 19 km^2 and 44 km^2, volumes of 76×10^6 m^3 and 230×10^6 m^3, and 110,000 Ci and 39,000 Ci, respectively).

The area near Mayak currently contains ~800×10^6 Ci of radioactive waste in various forms, which is clearly a serious environmental hazard, primarily because of the possible outflow of radionuclides into the Techa-Iset'-Tobol-Irtysh-Ob stream system that drains into the Kara Sea.

Since 1987, in order to reduce the hazard of recurring radiation accidents and disasters, Mayak has transformed part of the HLW into phosphate glass in an EP-500/l electrical furnace. About 250×10^6 Ci of HLW was vitrified by the end of 1995, including ~70×10^6 Ci of waste previously accumulated. In order to dispose of high-salinity HLW, a technology for its fractionation was developed, which allowed separation, concentration, and vitrification of long-lived fission products, while the major volume of waste was transformed to MLW, cemented, and bituminized.

Among new environmental problems at Mayak are storage and utilization of weapon-grade plutonium that is no longer needed for weapons because of disarmament agreements. According to Yablokov [14], about 250 t of plutonium were produced in the world for military purposes (in the USSR, 140 ± 25 t; U.S.A., 97± 8 t; Great Britain, 2.8 ± 0.7 t; France, 6 ± 1.5 t; and China, 2.5 ± 1.5 t). According to Seaborg [15], up to 60 t of plutonium were produced annually in the world as a result of the activity of nuclear power stations, and ~600 t of plutonium were accumulated in the spent fuel elements, and ~100 t of plutonium were recovered. From 1990 to 2050, nuclear power plants in the world are predicted to yield ~4800 t of high-background plutonium (^{239}Pu, ^{240}Pu, and ^{242}Pu); 30 t of ^{241}Am and 55.2 t of ^{243}Am; 4.8 t of ^{242}Cm and 18.6 t of ^{244}Cm; and ~ 270 t of ^{237}Np. Two approaches were proposed to address the problems of plutonium [15]: vitrification with other fission products and the safe burial or use of plutonium as uranium-plutonium fuel in nuclear power plants. In the latter case, in the 60-year period considered, the amount of long-lived isotopes would decrease by 28% for plutonium, 45% for ^{241}Am, and 3% for ^{243}Am, and increase by 166% for ^{242}Cm and 187% for ^{244}Cm, while the total amount of ^{237}Np would not change. Only the potential use of a new type of reactor (cascade liquid-salt "scavenger" reactor), or of the thorium fuel cycle in a breeder reactor or in a subcritical reactor (see Chapter VII) would allow the burning of the accumulated plutonium and the complete elimination of this environmentally dangerous element.

Nowadays, the reactor division of Mayak operates two reactors producing radionuclides for both military and civilian purposes. Five uranium graphite reactors were shut down between 1987 and 1991. Production of the weapon-grade plutonium at Mayak ceased in 1987. The radiochemical plant operation started in 1977, and since then its staff has reprocessed spent fuel from different power reactors, as well as from transport and research reactors. During the operation period since 1977, 2380 tons of spent fuel were received from domestic and foreign power plants for reprocessing. The plant has a spent fuel storage pool and three chopping-dissolution process lines.

At Plant RT-1 the PUREX process (see Chapter I) with tributyl phosphate (TBP) in a light diluent is used for reprocessing of spent fuel from WWER-440 reactors. The end product of the uranium line is the uranyl nitrate hexahydrate melt with 2.0-2.4% enrichment of ^{235}U. The melt is passed to fabrication of fuel elements for RBMK-1000 and RBMK-1500 reactors so that the fuel cycle of WWER-440 reactors becomes closed for uranium.

Plutonium extracted from WWER-440 spent fuel is now stored in the form of dioxide. The variants of the return of this Pu into the fuel cycle are discussed below. The use of mixed oxide fuel (MOX) is under consideration. The HLW from Plant RT-1 is solidified by means of vitrification at an industrial facility, the first line of which was put into operation in 1991. This facility handles accumulated HLW from defense reprocessing and current waste from the Plant RT-1. The glass blocks produced are delivered to an interim monitored storage. The first line of the HLW partitioning facility of the RT-1 plant was put into operation in 1996. It is used to reduce the volume of liquid HLW to be vitrified so that both the vitrification process and the subsequent storage of the glass blocks becomes less expensive [16].

Two of Russia's other radiochemical centers producing weapon plutonium are the Mining and Chemical Plant (MCP), located in an underground facility 40 km from Krasnoyarsk, and the Siberian Chemical Plant (SCP), located near Seversk (Tomsk-7), 20 km from Tomsk. There are several sources of radioactive pollution at these plants. At MCP, these are industrial nuclear reactors cooled with Yenisei River water, a radiochemical factory processing irradiated uranium blocks, and set-ups for the processing and storage of active waste. The Severnyi site for the disposal of liquid MLW is located 60 km from Krasnoyarsk, downstream on the Yenisei River. Radioactive waste (10^8 Ci) is injected into a sand-clay strata, resulting in the pollution of the environment in the adjacent territory and the entire Yenisei basin [17, 18].

The Siberian Chemical Plant is just as dangerous in terms of possible radioactive emission. It also included plutonium production, a radiochemical factory, and a factory for isotope separation. Russia's largest area for the disposal of liquid radioactive waste is located there in deep geological strata (~10^9 Ci was injected at this site). An explosion of a technological apparatus in the SCP in 1993 resulted in the release of ~10^3 Ci of radioactive products into the atmosphere (radionuclides of ^{95}Zr, ^{95}Nb, ^{106}Ru, and trace amounts of plutonium), including fallout of 500 - 600 Ci from the industrial zone of the plant.

Thus, all three of the radiochemical plants, Mayak, MCP, and SCP, are dangerous sources of radioactive contamination of the environment. Reliable and automated monitoring of the radiation in these regions and considerable work on the rehabilitation of territories contaminated with radionuclides are urgently needed.

4. Russian R&D on Recovery of Radionuclides from Radioactive Wastes

In order to achieve a closed fuel cycle, proper management of long-lived radionuclides contained in spent nuclear fuel, optimization of existing HLW reprocessing methods, and separation technologies for operating and new reprocessing plants have been studied during the last 20 years. These studies are carried out at several Russian institutes (V.G. Khlopin Radium Institute, Institute of Chemical Technology, Institute of Physical Chemistry, Institute of Geochemistry and Analytical Chemistry, and others) in collaboration with the radiochemical plants of Mayak and the Mining Chemical Association. Key operation of technologies for separation of long-lived radionuclides involves selective recovery from HLW of cesium, strontium, technetium, rare-earth, and transplutonium elements, as well as residues of uranium, neptunium, and plutonium remaining after the PUREX process.

4.1. Recovery of cesium and strontium using chlorinated cobalt dicarbollide (ChCoDiC process)

Fundamental studies on the extraction of Cs and Sr by cobalt (III) dicarbollide have been carried out primarily in the Czech Republic and are summarized in Chapters I and V. In Russia, dicarbollide technology has been applied to HLW processing (see Chapter V).

Figure 3-2 shows the nitric acid dependencies for the extraction of Sr and Cs using a 0.01M solution of cobalt(III) dicarbollide and 0.01M polyethylene glycol (PEG-400) in nitrobenzene. The flowsheet shows the combined recovery of cesium and strontium, as well as the extraction of barium and lead from HLW with HNO_3. The resulting content has not more than 3.0 M and the overall content of nitrate-ion of not more than 5.0 M. The combined stripping of cesium and strontium is carried out by means of 5.0-6.0 M HNO_3; the extractant is regenerated by a solution of 8 M/L HNO_3, and 20 g/L hydrazine, which is then used for the stripping operation. This extraction mixture recovers more than 99.5% of the cesium and strontium.

The most notable achievement in the use of the ChCoDiC process in Russia is now related to the reprocessing of HLW with different composition at Mayak PA. Using this technology, the first commercial facility in the world (UE-35) for the recovery of radionuclides began operation in August 1996. The first line of this facility was planned for the selective recovery of cesium and strontium from HLW. The results of this facility's operation for the first three months are:

- 320 m^3 of HLW with an activity of about 10 MCi were reprocessed;

- 99.96% of cesium was recovered and 99.94% of strontium;

- Glass blocks with specific activity up to 550 Ci/kg were produced;

- Cost of the glass production was decreased by 60%, and cost of separation was increased by 5%.

The raffinate produced after cesium and strontium recovery is used for separation of long-lived actinides.

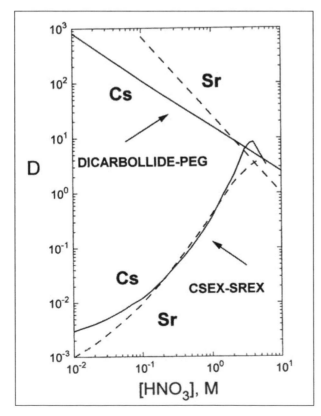

Figure 3-2.
Nitric acid dependences of Cs and Sr extraction (D is the distribution coefficient) using dicarbollide and crown ether extractants [19]. Dicarbollide (0.01M cobalt(III) dicarbollide and 0.01M PEG – 400 in nitrobenzene), CSEX-SREX (0.1M Cs extractant – 0.005M Sr extractant – 1.2M TBP – Isopar-L – 5 vol.% lauronitrile).

4.2. Recovery of actinides using neutral organophosphorus compounds

In Russia, neutral bifunctional extractants have been studied for a number of years. Parallel investigations were carried out in the U.S.A. (see Chapter I). To extract TRUs from HLW, Russian chemists have adopted a different carbamoylmethylphosphine oxide (CMPO) derivative than the one used in the U.S. TRUEX process, namely, diphenyl-N,N-di-n-butyl CMPO, which is abbreviated DPhDBCMPO. The diphenyl CMPO derivative is insufficiently soluble in paraffinic hydrocarbon diluents, even in the presence of excess TBP, to be of practical use. Furthermore, diphenyl CMPO derivatives have a strong propensity toward third phase formation. However, Russian chemists found that use of a fluoroether called Fluoropol-732 as a dilutent for the diphenyl CMPO eliminated the unfavorable solubility and third phase formation properties of this derivative. The D_{Am} versus aqueous HNO_3 concentration curve using a 0.05M DPhDBCMPO solution in Fluoropol-732 is shown in Figure 3-3. These data show that the values of D_{Am} obtained with the DPhDBCMPO-Fluoropol system are significantly higher, using only one-fourth the concentration of CMPO, than those obtained with the TRUEX process solvent over the entire nitric acid concentration range.

The Russian TRU extraction process uses a 0.1M solution of DPhDBCMPO in Fluoropol-732 as the process solvent (Figure 3-4). An 18-stage bank of centrifugal contactors was used to test the TRU extraction. Feed solution consisted of a HLW simulant, 5 M in HNO_3, containing more than 13 g/L of lanthanides and actinides. An interesting feature of the flowsheet is the use of acetohydroxamic acid (AHA) to strip Fe(III), Zr(IV)

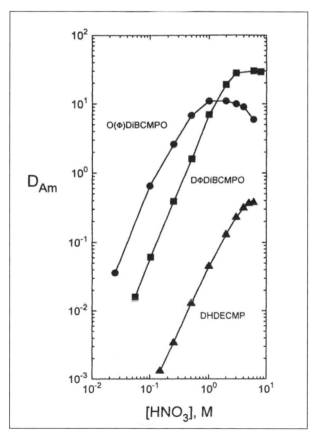

Figure 3-3.
Comparison of octyl(phenyl)- and diphenyl- DiBuCMPO and DHDECMP in the presence of TBP at 25°C as extractants for Am(III) in nitric acid [19]. 0.25M CMP or CMPO – 0.75M TBP – CCl_4.

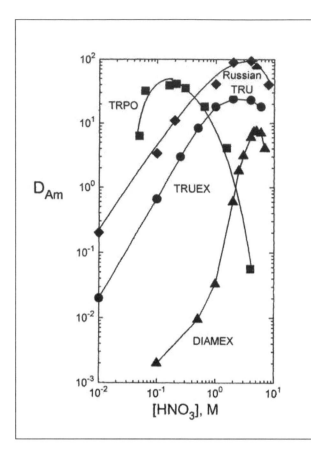

Figure 3-4.
Comparison of TRUEX, Russian TRU, DIAMEX, and TRPO process solvents as extractants for Am(III) in nitric acid [19]. TRUEX process solvent (0.02M O(∅)DiBCMPO - 1.4M TBP - Conoco (C_{12}-C_{14})), t=30°C. Russian TRU process solvent (0.05M D (∅) DBCMPO - – 1.4M TBP – Conoco (C_{12}-C_{14})), t=30°C. Russian TRU process solvent (0.05M D (Æ) DBCMPO – Fluoropol-732), t=23°C. DIAMEX process solvent (0.5M DMDBTDMA-TPH), t=25°C. TRPO process solvent (30 vol.% TRPO - kerosene), t=25°C.

and Mo(VI), which also extracts the transplutonium elements (TPE). A solution of 2 M HNO_3, 10 g/L AHA was employed for this purpose. The AHA strip solution is contacted with fresh process solvent to remove any traces of TPE and possibly Pu. TPEs and lanthanides were stripped from the process solvent using 0.01 M HNO_3. More than 99.5% of the actinides and lanthanides were recovered and concentrated by a factor of four to six. The reduction of Fe, Zr, and Mo from the TPE fraction was >50%. Efforts are currently underway to apply the process to a plant-scale operation.

A number of the favorable features outlined for TRUEX also apply to the Russian TRU process, namely, efficient extraction of Am(III) over a wide range of HNO_3 concentrations and stripping with low concentrations of acid. The Russian TRU process has the added advantage of using a lower concentration of a less expensive extractant. Because of the absence of TBP in the DPhDBCMPO-Fluoropol system, radiolytic and hydrolytic degradation is probably less than with the TRUEX process solvent.

The technology of the modified TRUEX process was confirmed by recent tests under static conditions and for industrial HLW from the Mayak facility.

4.3. Separation of actinides and lanthanides

4.3.1. Separation by partition countercurrent chromatography. The most effective method for TPE separation from highly radioactive wastes is extraction based on the use of bidentate neutral organophosphorus compounds (BNOC) (see, e.g., [19]). A technological scheme of TPE removal with diphenyl (dibutyl-carbamoylmethyl) phosphine oxide (Ph_2Bu_2) in fluorine-containing polar solvent using a 18-step set of extractors has been proposed. The possibility to isolate >99.5% TPE as well as to purify the TPE from accompanying elements except lanthanides has been shown. The TPE + lanthanides fraction contains a low concentration of nitric acid (about 0.02 - 0.03 M) that makes it possible to choose other water systems for group separation of these elements. It is possible to increase the separation factors for the elements in the BNOC-based systems by: 1) improvement of reagent selectivity by changing the reagent structure mainly by introduction of "hard" bridge fragments (arresting the arrangement of the donor atoms) into the extractant molecule, 2) addition of complex-forming agents to the aqueous phase that are able to react selectively with TPE or lanthanides in acid solutions. However, neither the first nor the second method allows single stage separation of TPE from lanthanides as they have very similar properties. A multistage liquid-liquid extraction separation is needed to separate the TPE and lanthanides in order to increase the separation factors.

Separation of actinides and lanthanides can be carried out by partition countercurrent chromatography (CCC), which is also called liquid chromatography with a free stationary phase. CCC is a relatively new method of separation. This method, suggested by the American scientist Ito, is based on the retention of the stationary organic phase in a rotating column under the action of centrifugal forces while the mobile aqueous phase is continuously pumped through. Substances are separated due to differences in their distribution coefficients when the mixture to be separated moves with the flow of the aqueous phase through the column. It was shown that TPE and lanthanides can be separated in various BNOC-based systems using CCC [20].

4.3.2. Separation of actinides and lanthanides by precipitation. The search for new complex-forming and oxidizing systems for TPE and REE has opened up new possibilities in the chemistry and technology of these elements. Ferrocyanide ions are known to form a multitude of salts with, in particular, cations of various elements and rare earths. The formation of salts with the same cations occurs also with ferricyanide ions but to a lesser degree. Use of the ferri/ferrocyanide system has been described in various redox reactions. Study of the behavior of some TPE and REE in weakly acid and alkaline solutions containing ions of tri- and bivalent iron is of immediate interest.

When potassium ferrocyanide is added to an americium solution in 0.1 M nitric acid, a white precipitate forms, just as in the case of REE. However, when potassium ferricianide is added, an americium species precipitates from the acidic solution, contrary to REE (Table 3-4). TPEs and REEs are separated at the stage of TPE ferricyanide precipitation. In this case, REEs remain in acidic solution and can be easily separated from precipitated TPEs. Americium is isolated from curium at the stage of its alkaline-ferricyanide dissolution [21].

5. Environmental Monitoring in the Mayak Area

The long-term radiation hazard to contaminated territories arises most notably from the actinide isotopes. Removal of these radionuclides from ground water and surface water basins, soils, and other natural and manmade media is the most important, yet most difficult problem.

TABLE 3-4. Precipitation of Am (III) and Eu (III) Taken Separately from Nitric Acid Solutions Containing Potassium Ferro- and Ferricyanide

Precipitant	Element	Taken		Found in the sediment %	Found in the solution %
		M	Mg/mL		
[K$_4$Fe(CN)$_6$]	Am(III)	4.1×10^{-3}	1.00	99	1
	Eu(III)	8.6×10^{-3}	1.31	98	2
[K$_3$Fe(CN)$_6$]	Am(III)	6.4×10^{-3}	1.55	92	8
	Eu(III)	8.6×10^{-3}	1.31	0	100

Estimation of the efficiency of expensive remediation measures depends on knowledge of the dynamics of radionuclide migration in surface and ground water. Some institutes of the Russian Academy of Sciences are studying interactions between ground water and bedrock on the basis of the chemical and mineralogical composition of the samples taken from various sites of this zone and from various depths using boreholes. We have performed radiochemical analysis of liquid and solid samples and determined the content and the forms of occurrence of radionuclides. Prediction of the migration process is a very complex scientific problem even for small areas because of the diversity of chemical and mineralogical compositions of environmental matrices.

Radionuclide content in rocks and water has been determined in accordance with a procedure developed at the Vernadsky Institute [22-26]. This procedure makes determining several radionuclides from the same sample possible, which saves labor time spent on sample preparation and analysis. Analysis of soils, rocks, and other solid samples includes air drying, disintegrating (milling), sieving through 1mm filter pores, and igniting at a temperature of approximately 550°C to destroy the organic matter, all of which takes several days. Radionuclides from water samples have been concentrated by evaporation (at the Vernadsky Institute) and ultrafiltration (at the central Mayak plant laboratory).

During the first stage of the applied analytical scheme prior to radiochemical analysis, gamma spectrometry using a Ge(Li) detector is performed. Determination of Sr and TRU, which are pure b and a-emitters, can be done only radiometrically with radionuclide separation from a large sample of complicated chemical and radiochemical composition. This procedure needs careful purification, especially during the separation of radionuclides having almost similar energies of b- and a-emission. This problem relates to natural radionuclides, in particular those with amounts in the environment that are 2 to 3 orders of magnitude higher than those of artificial radionuclides. To transfer radionuclides from ashed residue or water concentrates, treating the samples with 7-8 M HNO$_3$ in the presence of potassium bromate after pretreatment by a mixture of HF and H$_2$SO$_4$ is sufficient. Further plutonium concentration and radiochemical separation can be carried out on the anionite VP-1Ap; for americium and curium, the complexation sorbent polyarsenazo-n has been used; and for Sr, the porous copolymer of styrol with divinylbenzole (TVEKS), impregnated with 10% dicyclohexyl-18-crown-6 (DCH18C6) in tetrachloroethane. This method is selective for ^{90}Sr, but needs careful separation from ^{210}Pb if the latter is present in comparable quantities. Therefore, the Mayak samples needed additional purification by preliminary elution of the radioactive ^{210}Pb by sodium oxalate solution. Radioactivity of the separated TRU has been measured at an alpha-spectrometer facility consisting of an ionization camera with a grid and an AI-4096 analyzer.

Extraction and determination of neptunium is more complicated than extraction and determination of plutonium and americium due to the generally lower concentrations of neptunium. The method developed at the Vernadsky Institute extracts neptunium using potassium phosphor tungstate during sorption on a porous teflon membrane impregnated with 0.5M solution of trioctylammonium nitrate (TOMAN) in toluene, which is followed by luminescence determination. $PbMoO_4$ has been used as a crystallophosphor [25].

Figure 3-5 shows luminescence spectra of crystallophosphorus based on $PbMoO_4$ and $NaBi(WO_4)_2$ and activated by neptunium, plutonium, and americium. The intensity of luminescence (proportional to the concentration of TRU) is measured in the IR part of the spectrum, where the spectral lines of other elements are absent. Thus, the luminescence method is fairly selective and valid for the combined determination of neptunium, plutonium, and, in some cases, uranium in a single sample. The method is used for the analysis of uranium, neptunium, and plutonium in fallout solutions from the radiochemical industry and natural objects without preliminary separation of analyzed elements and irrespective of their isotopic composition. Plutonium can be determined in samples containing an active nuclide such as americium. Thus, this method solves a problem which cannot be solved with methods currently in use. At the Vernadsky Institute, a simple filter photometer was produced for luminescence analysis of actinides in environmental objects and for ecological monitoring in regions near nuclear power and radiochemical plants.

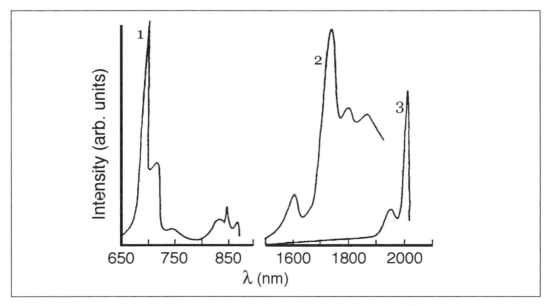

Figure 3-5. Luminescence spectra of cristallophosphorus $NaBi(WO_4)_2$ contaminated with Am (1) and of $PbMoO_4$ contaminated with Np (2) Pu (3).

To determine the chemical yield of plutonium, americium, and neptunium, traces of ^{236}Pu, ^{243}Am, and ^{239}Np are inserted into a solid sample before its ignition or into a water sample before its concentration. Detection limits of radionuclide determination under the above mentioned procedures are shown in Table 3-5.

To predict the migration of radionuclides and to develop remediation approaches, the content of radionuclides in particular components of biogeocenoses as well as their occurrence forms must be determined. As for speciation of radionuclides in solid materials, determining geochemical forms of mobility seems most appropriate [26]. These data are usually obtained by selective leaching. For loamy materials, which mostly accumulate radionuclides, the fraction of relatively mobile forms is a maximum for radiostrontium, whereas for plutonium it usually does not exceed 10%. This fact explains the high concentration of ^{90}Sr in ground water from various sites of the region as compared to plutonium and cesium. The mobility of strontium is higher than that of its natural analogs, calcium and magnesium. Our results show that the migration ability of radionuclides in the ecosystems in question increases in the following series:

$$^{137}Cs < {}^{239,240}Pu < {}^{241}Am < {}^{90}Sr < {}^{60}Co < {}^{237}Np.$$

TABLE 3-5. Radionuclide Relative Detection Limits Using Described Procedures

Radionuclide	Detection limit [Bq/g]
Co-60	$1.2*10^{-3}$
Sr-90	$4.5*10^{-3}$
Np-237	$2.6*10^{-6}$
Pu-239	$1.3*10^{-5}$
Am-241	$2.6*10^{-5}$

The data on the distribution of radionuclides ^{90}Sr, ^{137}Cs, ^{237}Np, ^{239}Pu, and ^{241}Am among components of various ecosystems of the Southern Ural region as well as their forms of occurrence show that ^{90}Sr, ^{237}Np, and ^{241}Am are mostly contained in compounds of fulvic acids, which is the reason that they have such a high mobility in the environment. On the contrary, considerable amounts of ^{137}Cs and ^{239}Pu have been found in low soluble humic acids bonded primarily with calcium and relatively low-mobile hydroxides. Data on vertical migration of plutonium, radiostrontium, and radiocesium in various types of soils have been obtained and corresponding coefficients have been calculated. The content and nature of organic substances is considered to be one of the most important factors affecting the migration resistance of soil media. In particular, the dependence between diffusional resistance of upper soil layers to plutonium and radiostrontium mass transfer and the content of humus in these layers has been determined. The correlation between migration coefficients and the content of the most mobile plutonium forms has also been determined.

6. Conclusion

The most important problems of modern radiochemistry as applied to current environmental problems are summarized below.

Radioactive waste management:

• Creation of new processes of spent nuclear fuel reprocessing;

- Partitioning of the accumulated radioactive waste;

- Development of new approaches to ensuring long-term ecologically safe radionuclide storage methods (e.g., repository storage);

- Development of new matrices and new methods of vitrification of radionuclides with various half lives; and

- Development of new types of barriers (natural and manmade) for the safe storage of spent nuclear fuel and radioactive wastes.

The behavior of radioactive substances in the biosphere:

- Development of principles and approaches of radiomonitoring in regions affected by nuclear plants and research and production facilities;

- Elaboration of new methods and techniques of radioecological analyses of environmental samples;

- Data gathering, systematization, and computation on content and speciation of radionuclides in the environment; and

- Long-term forecasts of radionuclide behavior after their discharge to the environment.

References

1. Myasoedov, B.F. (1998) *Environmental Geoscience* **1**, 3.
2. Sivintsev, Yu.V. (1988) *At. Energy* **64**, 46.
3. UNSCEAR Reports on Ionizing Radiation to the General Assembly. United Nations, New York (1982, 1988, and 1993).
4. Makhan'ko, K.P (ed.) (1994) *Radiation Conditions in the Territories of Russia and Adjacent Countries in 1993*. Yearbook, NPO Taifun, Obninsk.
5. Pavlotskaya, F.I., Fedorova, Z.M., Emel'yanov, V.V., et al. (1985) *At. Energ.* **59**, 381.
6. Babaev N.S., Dermin, V.F., Il'in, L.A., et al. (1984) In *Nuclear Energy: Humans and the Environment*. Atomizdat, Moscow, 1.
7. Information on the Chernobyl NPP Accident and Its Consequences Prepared for IAEA, *At. Energ.* **61**, 301 (1986).
8. Izrael, Yu.A. (ed.) (1990) *Chernobyl: Radioactive Contamination of the Environment*. Hydrodrometeoizdat, Leningrad.
9. Health Hazard from Radiocesium following the Chernobyl Nuclear Accident. Report of the World Health Organization Working Group, *J. Envir. Radioact,* **10**, 257 (1989).
10. Fetisov, V.I. (1996) *Radiation Safety Problems,* **1**, 5.
11. Mokrov, Yu.G. (1996) *Radiation Safety Problems,* **3**, 19.
12. Romanov, G.N., Spirin, D.A., and Alexahin, P.M. (1990) *Priroda* **5**, 53.
13. Drozko, E.G., Ivanov, I.A., et al. (1996) *Radiation Safety Problems* **1**, 11.
14. Yablokov, A.V. (ed.) (1994) Plutonium in Russia. *Ecology, Economy and Politics*. Sots.- Ecol. Soyuz, Moscow.
15. Seaborg, G.T. (1995) Transuranium Elements: Past, Present, and Future, *Acc. Chem. Res.* **28**, 257.
16. Myasoedov, B.F., and Drozko E.G. (1998) *Journal Alloys and Compounds* **271/273**, 216.
17. Vakulovskii, S.M., Kryshev, I.I., Nikitin, A.I., et al. (1994) *Yad. Energ.* **2-3**, 124.
18. Nosov, A.V., Amanin, M.V., Ivanov, A.B., and Mar, A.M. (1993) *At. Energ.* **74**, 144.
19. Horwitz, E.P., and Schultz, W.W. (1999) Solvent Extraction in the Treatment of Acidic High-Level Liquid Waste. In *Metal Ion Separation and Preconcentration*. Amer. Ch. Soc.

20. Myasoedov, B.F., and Chmutova, M.K. *3rd Japan-Russian Joint Symposium on Analytical Chemistry*, November 5-9, 1986. Nagoya, Japan.

21. Kulyako, Yu.M., Malikov, D.A., Trofimov, T.A., and Myasoedov, B.F. (1996) *Mendeleev Commun.*, 173.

22. Myasoedov, B.F. (1994) *Journal of Alloys and Compounds* **213/214**, 290.

23. Kremlyakova, N.Yu., Novikov, A.P., Korpusov, S.G., and Myasoedov, B.F. (1991) Patent Rus., 4924649/25.

24. Novikov, A.P., Mikheeva, M.N., Trofimov, T.I., and Kulyako, Yu.M. (1990) Patent Rus., 4897664/23.

25. Novikov, A.P., Ivanova, S.A., Mikheeva, M.N., and Myasoedov, B.F. (1992) Patent Rus., 5146273/27.

26. Myasoedov, B.F., Novikov, A.P., and Pavlotskaya, F.I. (1996) *Zhurnal Analyticheskoy Khimii* **51**, 124.

Boris F. Myasoedov is Professor of Chemistry and Academician of the Russian Academy of Sciences (RAS). He has been Head of the Laboratory of Radiochemistry of the Vernadsky Institute of Geochemistry and Analytical Chemistry of the RAS since 1969. He received his Ph.D. in Chemistry from the Vernadsky Institute. His fields of expertise are methods and techniques of concentration, separation, isolation, and identification of radioactive nuclides and different toxicants using solvent extraction, sorption, spectrophotometry, electrochemistry, and other methods. He is the author of more than 480 articles and monographs. Professor Myasoedov is Deputy Secretary General for Science of the RAS, President of the Russian Scientific Council on Radiochemistry, and Vice-President of the Scientific Council on Analytical Chemistry of the RAS. He is Editor of the Journal *Radiokhimia* (Russia), and Associate Editor of the *Journal of Radioanalytical and Nuclear Chemistry Letters*. In 1986, he received the USSR State Prize for his research work on the chemistry of transplutonium elements and the Khlopin Prize of the Academy of Sciences of the USSR.

PART 2

NUCLEAR SEPARATION TECHNOLOGIES
Future Directions

CHAPTER IV

NON-AQUEOUS SEPARATION METHODS

GREGORY R. CHOPPIN
Department of Chemistry
Florida State University
Tallahassee, Florida 32306, U.S.A.

ABSTRACT

Traditional separation processes in nuclear fuel and environmental remediation systems have used aqueous processes. Non-aqueous processes, however, have been used for uranium isotope enrichment, electrorefining of plutonium metal, and production of metallic fuel for advanced nuclear reactors. These non-aqueous processes are based on differences in properties such as volatility of various compounds and redox thermodynamics of the actinide elements in molten salt media.

Application of non-aqueous processes to the treatment of radioactive waste is of growing interest. Non-aqueous processes, in general, have the advantages of providing higher radiation resistance, using more compact equipment, producing smaller amounts of secondary waste volume, and having higher proliferation resistance than aqueous processes. The disadvantages are greater difficulty in conducting the separations and smaller decontamination factors, in general, than aqueous processes.

Partitioning/transmutation (P/T) concepts for the destruction of radionuclides in nuclear wastes to reduce the long-term hazard of nuclear waste stored in repositories are also of interest. As it is necessary to process and recycle the irradiated material after relatively short cooling times, non-aqueous systems of higher radiation resistance have been considered for the primary treatment. Both fluoride volatility and molten salt systems are under consideration.

1. Introduction

Separation processes in nuclear fuel and environmental remediation systems typically have involved aqueous solutions. Non-aqueous processes have been extensively used for uranium isotope enrichment, electrorefining of plutonium metal, and production of metallic fuel for advanced nuclear reactors. Such non-aqueous processes are based on differences in properties such as volatility of the compounds being separated or redox thermodynamics of the actinide elements in molten salt media. At present, application of non-aqueous processes to the treatment of radioactive waste is also of renewed interest.

Advantages of non-aqueous processes include generally higher radiation resistance, more compact equipment, smaller amounts of secondary waste volume, and higher pro-liferation resistance than aqueous processes. The disadvantages of non-aqueous methods are the greater difficulty of conducting the separations and smaller decontamination fac-tors, in general, than aqueous processes. Also, some non-aqueous treatments are very sen-sitive to even small amounts of moisture or oxygen and must take place in isolated cells under an inert atmosphere.

2. Volatility Methods

The gaseous diffusion process using the volatility of uranium hexafluoride has been the standard method of uranium isotopic separation for half a century [1]. Volatility tech-niques with fluorides have been used also to separate uranium and plutonium in irradiated fuel [2] and were studied for use in fuel processing in the molten-salt reactor project at Oak Ridge National Laboratory, U.S.A. (Figure 4-1) [3]. The separation of uranium and plutonium from the fission products in irradiated nuclear fuel is limited in these processes because volatile fluorides are formed by several fission products (in particular, by iodine, technetium, and tellurium). As a non-aqueous process, a fluoride violatility process produces relatively small volumes of contaminated liquids; efficiently separates U and Pu by distillation; partitions Tc in a straightforward matter; and does not present solvent degradation problems.

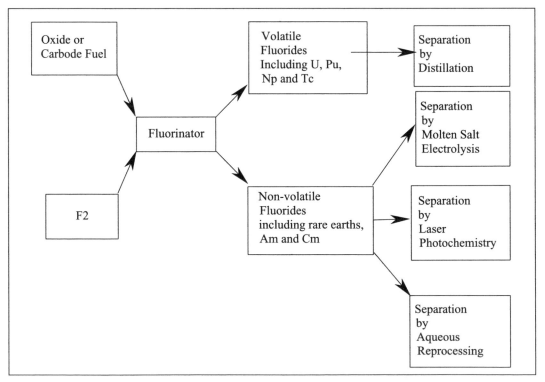

Figure 4-1. Conceptual flowsheet of a general fluoride volatility process.

Tellurium fluoride (Te_2F_{10}), technetium fluoride (TcF_6), and iodine fluoride (IF_5) have similar chemistry to that of UF_6 and usually are contaminants in the uranium and plutonium streams in the volatility process. Tellurium and iodine as the compounds TeF_6 and IF_7 have a much higher volatility than UF_6 and PuF_6, and, as a result, they can be separated after oxidation from uranium and plutonium. The removal of technetium contamination is a difficult task in the fluoride volatility process because the TcF_6 diffuses readily with the UF_6 and PuF_6 streams. The radiolytic decomposition of PuF_6 to PuF_4 and F_2 in the volatilization process is a disadvantage of the method since PuF_4 is less volatile and deposits on the surfaces of the equipment. Addition of fluorinating agents such as CIF_3 can prevent the decomposition, reforming PuF_6 and decontaminating the equipment of plutonium by volatilization of the PuF_6. A powerful fluorinating agent O_2F_2 ("foof") has been developed, which in pilot-plant tests has shown the ability to fluorinate PuO_2 to PuF_6 at ambient temperatures in a simple process [4].

Other volatility processes that could be of use in the separation of actinides and other radioactive elements are not as well developed as those that use fluoride volatility. In general, chloride volatility is less selective than fluoride volatility because of the larger number of volatile chlorides. However, the volatility of $ZrCl_4$ may be useful in removing the zirconium cladding from spent fuel elements [5]. A scheme has been proposed for transmutation of technetium to the nonradioactive ruthenium in which the product Ru is converted to RuO_4 by ozonolysis and separated from the remaining Tc_2O_9 by the higher volatility of RuO_4 [6]. Research is continuing on evaluation of these volatility processes for use in practical full-scale separation systems.

Several b-diketone ligands have been shown to form relatively volatile compounds with trivalent lanthanides. Based on the strong similarity in chemistry of the 4f and 5f elements, similar volatility can be expected for trivalent actinides. It is known that the tetravalent and hexavalent ions of uranium and plutonium form volatile compounds with b-diketonates such as "fod" (6,6,7,7,8,8,8-heptafluoro-2,2-dimethyl-3,5-octanedione), with large separation factors for uranium and plutonium from Am [5]. Fod is stable to oxidation by air and water at temperatures 100°C and can be recovered by extraction with an organic solvent; however, its radiation stability has not been reported. Extraction of b-diketonate complexes from large amounts of aqueous wastes presents engineering problems which must be solved before use of these volatile compounds could be considered a viable separation option for nuclear wastes.

3. Pyrochemical Processes

Several molten salt processes have been investigated for the treatment of spent fuel from molten salt breeder reactors, such as the Liquid Metal Fast Breeder Reactor (LMFBR) [7] and the Experimental Breeder Reactor II (EBR-II) [8], as well as for spent fuel from light water reactors (LWRs) [5]. The inherent radiation resistance of molten salts, which allows the processing of spent fuel after very short cooling periods, is a major advantage of molten inorganic salts in spent fuel reprocessing. Moreover, the decay heat of "fresh" reactor fuels can be used to maintain the necessary process temperatures of 500–800°C of the molten salt fluids.

In the Molten Salt Reactor Experiment (MSRE), carried out in the 1960s at Oak Ridge National Laboratory, a mixed fluoride salt was used as coolant, fuel, and blanket system. The working fluid consisted of BeF_2, ^7LiF, ThF_4 and UF_4, and provided a fertile thorium blanket, a neutron multiplier, the coolant for the fuel and the reactor, and a reprocessing solvent [5]. Actinides and fission products from the working molten salt were removed by molten metal extraction and fluoride volatility processes. A major problem of the process was the severe corrosion of equipment, especially if moisture invaded the system.

Electrorefining is the electrochemical procedure that had been proposed for the purification of plutonium metal alloys as fuel for the Los Alamos Molten Plutonium Reactor Experiment (LAMPRE) [5]. LAMPRE never went into operation, but electrorefining was adapted to the purification of plutonium and uranium metals for both the weapons and breeder reactor fuel programs. The separation technique was based on a sequential oxidation process from a molten pool of feed metal acting as the anode. The cations would move through an ionic molten salt to the cathode where they were reduced and deposited. The process is somewhat similar to the Salt Transport Process; the driving force, however, is provided electrically and can be controlled very precisely to obtain cathode products of high purity.

The Salt Transport Process [7,9] was designed to recover plutonium and uranium from spent fuel from fast reactors. The separation process is based on the transport of actinides between two molten alloys of magnesium, one containing copper, the other zinc. The two alloys are connected by an ionic molten salt solvent that allows the migration of ions between the alloys. This ion transport is driven by the difference in the thermodynamic activities of plutonium and uranium in the two alloys. The process seems to be applicable for the reprocessing of metal fuels used in fast reactors with a high burn-up and a short decay cooling time.

Argonne National Laboratory has developed an electrometallurgical separation (EMT) scheme for the reprocessing of spent fuel from the Experimental Breeder Reactor (EBR-II) based on a molten salt electrochemical system (Figure 4-2) [8]. In the "direct transport"

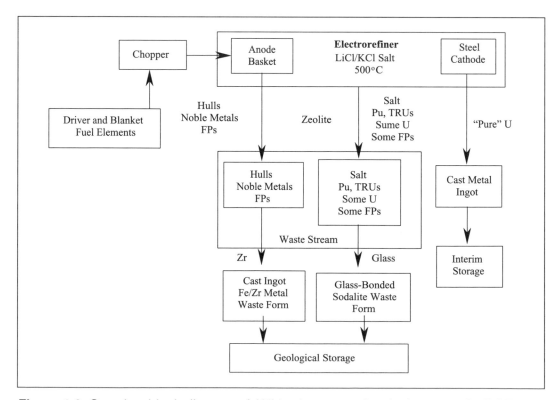

Figure 4-2. Overview block diagram of ANL's electrometallurgical process for DOE spent nuclear fuel.

process, uranium fuel is anodically dissolved as U^{3+}, from a pool of molten cadmium into a molten salt and transported through the salt by forced convection to a cathode where it is deposited as metallic uranium. In an alternative deposition process, dissolved U^{3+} is transported through the salt by forced convection to a cadmium pool where it is reduced to the metal. After reversing the polarity of the cadmium pool, uranium is reoxidixed and dissolved in the salt from which it is deposited on a steel cathode. The process produces two high-level waste forms, a zeolite containing transuranic elements and some fission products, and a metallic waste form containing cladding, noble metals, and fission products. Spent fuel from the Experimental Breeder Reactor (EBRII) has been successfully reprocessed by this electrometallugical method. The latest model of the electrorefining apparatus in which the Cd pool is eliminated is shown in Figure 4-3.

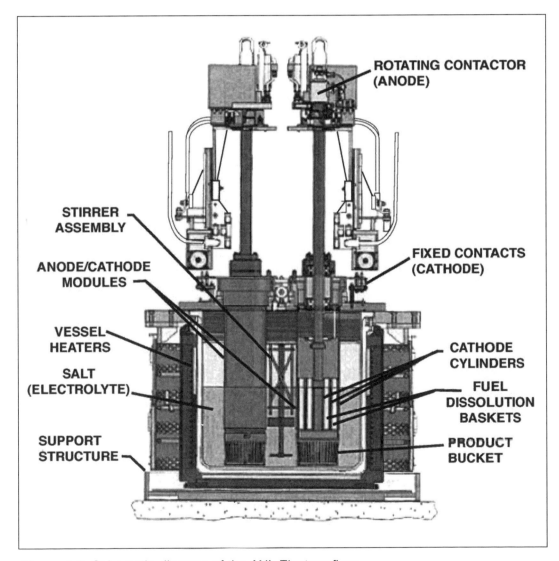

Figure 4-3. Schematic diagram of the ANL Electrorefiner.

An interesting possibility for the use of molten salt systems for the remediation of alkaline tank wastes involves direct extraction of actinides from molten nitrate media with tributyl-phosphate (TBP). Studies of the distribution coefficients by TBP of several actinides in extraction from a $LiNO_3+KNO_3$ eutectic mixture gave extraction efficiencies larger than those for extraction from an aqueous phase [9]. More research is necessary to assess the applicability of these systems to radioactive wastes.

4. Separations for Partitioning and Transmutation

The Partitioning and Transmutation (P/T) concept whereby the longest-lived radionuclides in nuclear wastes would be partitioned and transmuted by irradiation in an accelerator or a nuclear reactor to new nuclides of much shorter half-lives is of considerable international interest. This approach could decrease the concern of having to ensure placement of this waste in permanent ($>10^4$y) storage.

Several P/T concepts for the destruction of radionuclides in nuclear wastes are presently under consideration [4, 10, 11]. All rely on highly effective separation processes in order to obtain pure radionuclide fractions that can be used to prepare the material to be transmuted. That material would be irradiated with high neutron fluxes in order to achieve transmutation or fissioning into radioelements with sufficiently shorter half-lives. A variety of separation processes, both aqueous and non-aqueous, have been proposed for the various transmutation concepts.

Transmutation in LWRs requires initial separation of uranium and plutonium from minor transuranic elements, fission products, and zircalloy cladding hulls of the spent nuclear fuel. A PUREX-type process [12] could be a suitable first separation step, but might be inadequate to achieve the high separation factors necessary for P/T. Subsequent processes [12] such as TRUEX or DIAMEX would have to be included in the total system. The separated TRUs, Tc, and I would be prepared for transmutation. The majority of separation processes that have been proposed recently for the processing of neutron-irradiated materials are nonaqueous, including fluoride volatility processes and molten salt systems. This is due to the necessity for multiple irradiations to achieve the necessary high degree of transmutation. The irradiated materials must be processed between irradiations to isolate the remaining long-lived nuclides for the next irradiation cycle.

These separation processes are particularly challenging because the short cooling periods between irradiation and product separation result in intense radiation fields. The fact that actinides, technetium, and iodine must be processed separately requires the handling of a large number of process streams. The partitioning/transmutation concepts currently under consideration may become important components of the management of nuclear wastes but would require a major effort in improving and developing non-aqueous separation processes.

5. Conclusion

New processing plants are very likely to use non-aqueous separation methods because of their advantages over aqueous processes. In countries with large amounts of contaminated wastes from the defense work of the Cold War, these techniques may also be used in specific decontamination and remediation treatments.

Preparation of this chapter was supported by a contract from the U.S. DOE-BES Division of Chemical Science.

References

1. Villani, S. (1979) *Uranium Enrichment.* Springer Verlag, New York.
2. Hyman, H. H., Vogel, R. C., and Katz, J. J.(1956) Fundamental Chemistry of Uranium Hexafluoride Distillation Processes for the Decontamination of Irradiated Reactor Fuels. In *Progress in Nuclear Energy*, Series III, Process Chemistry, vol.1, Bruce, F. R., Fletcher, I. M., Hyman, H. H., and Katz, K. J. (eds) McGraw-Hill Book Co., Inc., New York.
3. Rosenthal, M. W., Haubenreich, P. N., and Briggs, R. B. (1972) *The Development Status of Molten Salt Breeder Reactors*, Report ORNL-4812, Oak Ridge National Laboratory.
4. OECD (1998) *Status and Assessment Report on Actinide and Fission Product Partitioning and Transmutation*, OECD Report NEA/PTS/DOC(98)6.
5. National Research Council (1995) *Nuclear Wastes: Technologies for Separations and Transmutation*, National Academy Press, Washington, D. C.
6. Dewey, H. J., Jarvinen, G. D., Marsh, S. F., Marsh, N. C., Smith, B. F., Villareal, R., Walker, R. B., Yarbro, S. L., and Yates, M. A. (1993) *Status of Development of Actinide Blanket Processing Flowsheets for Accelerator Transmutation of Nuclear Waste*, Report LA-UR-93-2944, Los Alamos National Laboratory.
7. Knighton, J.B., and Baldwin, C.E. (1979) *Pyrochemical Processing of UO$_2$-PuO$_2$ LMFBR Fuel by the Salt Transport Method*, Rocky Flats Report. RFP-2887, CONF-790415-29.
8. National Research Council (2000*) Electrometallurgical Techniques for DOE Spent Fuel Treatment*, Final Report, National Academy Press, Washington, D. C.
9. Christensen, D. C., and Mullins, L. J. (1983) Plutonium Metal Production and Purification at Los Alamos. In Carnall,W. T., and Choppin, G. R. (eds.), *Plutonium Chemistry*, A.C.S. Sym. Ser. #216, Am. Chem. Soc., Washington, D. C., 409.
10. Isaac, N.M, Fields, P.R., and Gruen, D.M. (1961) *J. Inorg. Nucl. Chem.* **21**, 152.
11. *A Roadmap for Developing Accelerator Transmutation of Waste (ATW) Technology*, A Report to Congress, DOE/RW-0519, U.S. DOE, Washington, D. C. (1999).
12. Choppin, G.R., and Morgenstern, A. (2000) *J. Radioanal. Nucl. Chem.* **243**, 45.

Gregory R. Choppin is R. O. Lawton Distinguished Professor of Chemistry at Florida State University (FSU), Tallahassee, Florida, U.S.A. From 1968 to 1977, he was Chair of the Chemistry Department at FSU. He was a member of the U.S. National Research Council—National Academy of Sciences Board of Chemical Science and Technology, and is a member of the Board of Radioactive Waste Management, the Committee on Electrometallurgical Processing of DOE Spent Nuclear Fuel, the Committee on Remediation of Buried and Tank Wastes, and the Committee on a Long-Term Environmental Quality R&D Program in the U.S. DOE. He received his B.S. degree from Loyola University (New Orleans, LA), a Ph.D. in chemistry from University of Texas (Austin, TX), and Honorary D.Sc. degrees from Loyola University and from Chalmers University of Technology (Sweden). He has received awards from the American Institute of Chemistry, the American Chemical Society, the American Nuclear Society, and the British Royal Society of Chemistry.

CHAPTER V

U.S.–RUSSIAN COOPERATIVE PROGRAM IN RESEARCH AND DEVELOPMENT OF CHEMICAL SEPARATION TECHNOLOGIES

VALERY N. ROMANOVSKY
V.G. Khlopin Radium Institute
St. Petersburg, Russia

ABSTRACT

The urgency of developing adequate separation technologies for the management of long-lived radionuclides is discussed in this paper. Separation of high-level waste (HLW) is needed to transmute long-lived radionuclides and to prepare extremely durable matrices for geological disposal. In reprocessing liquid HLW, particularly accumulated defense HLW, long-lived radionuclides (Cs, Sr, Tc, actinides, and rare-earth elements) can be recovered and subsequently converted into small volumes for solidification. This reduces considerably the cost of the solidification, storage, and disposal.

The results of the most advanced Russian research and development (R&D) on extraction technologies, especially with the use of phosphine oxides and chlorinated cobalt dicarbollide (ChCoDiC), are presented. These processes, based on different-radical phosphine oxide (POR) and ChCoDiC, were developed jointly by the Khlopin Radium Institute (KRI) and the Idaho National Environmental and Engineering Laboratory (INEEL) and were successfully tested on actual HLW at INEEL.

The results of the operation of the HLW partitioning facility (UE-35) at the Mayak facility are described. Since 1996, more than 1000 m^3 of defense waste have been reprocessed in four campaigns, and about 15 MCi of Cs and Sr were recovered. Now the technology is being developed for a second line of this facility in order to recover actinides and rare-earth elements along with Cs and Sr.

Data are presented on the development by KRI and INEEL of a universal extraction process (UNEX process) for simultaneous recovery of Cs, Sr, and actinide and rare-earth elements from HLW. In a series of joint tests of the UNEX process with actual INEEL HLW (sodium-bearing wastes and calcine solutions), efficiencies in the recovery of long-lived radionuclides were attained that permit conversion of the main bulk of HLW into Class A low-level waste (LLW). The UNEX process is being considered at INEEL for reprocessing sodium-bearing wastes and calcines and at Mayak for the second line of the UE-35 facility.

1. Introduction

Transmutation is a promising method for the management of long-lived radionuclides (see, e.g., Chapter VII). Another promising method is creation of durable matrices needed for disposal into geological formations. The technology for production of synthetic materials for such matrices is now under development in Russia with the use of high-temperature synthesis of minerals of the zircon group, garnet, cubic zirconium oxide, and others [1]. The principal feature of these minerals consists of their ability to include into the mineral matrix an individual element or at least chemical analogs. Such compositions result in materials very close to natural minerals known to have been stable for millions of years. These minerals, however, cannot incorporate the unseparated mixture of nuclides contained in spent nuclear fuel (SNF). Thus, for both transmutation and the synthesis of highly retentive matrices for geological disposal, selective recovery of long-lived radionuclides contained in SNF is an essential need.

The present worldwide practice for the management of long-lived radionuclides consists of vitrification of a non-separated mixture of long-lived radionuclides, and transportation of the resultant glass blocks to interim monitored storage in a special storage facility. However, even the presently practiced management of radioactive wastes requires selective recovery of long-lived radionuclides because the accumulated high-level wastes have a very complicated composition and contain, as a rule, large amounts of salts. In principle, such an unseparated mixture of radioactive waste (radwaste) could be solidified without any preliminary treatment, but in that case the following drawbacks inevitably emerge:

- large volumes of ballast material need to be vitrified along with radionuclides, which increases the cost at the stage of glass boiling, as well as at the stage of monitored storage of the glass blocks;

- many macrocomponents of non-separated radwastes have a harmful effect on the glass boiling process, its safety, and the quality of the glass blocks.

Because of these problems, an alternative method for HLW management by solidification is preferable, which would have the following results:

- long-lived radionuclides are separated from the HLW bulk and are concentrated in small volumes of solutions to be solidified, particularly by vitrification for interim monitored storage and subsequent disposal;

- the ballast mass of radwastes that remains after recovery of the radionuclides is solidified by cementing for near-surface storage as low-level waste (LLW).

The necessary degree of recovery of long-lived radionuclides is specified by the limiting standards imposed on their concentrations in LLW for near-surface storage. Attainment of the standard specifications provides a significant economic advantage. For example, specialists in the United States have calculated that separation of radionuclides from Hanford HLW would require 12,000 containers for vitrified blocks instead of 40,000 needed for wastes without separation; in the case of separation, the cost saving would be around U.S. $14 billion. Separation of radionuclides from HLW offers additional economic savings at different stages of management. Data indicate that the production cost of HLW in oxide form is $2,126 /kg compared with $64 /kg for LLW.

The need for efficient technologies for radwaste management can be summarized with the following arguments:

- the innovative technique for the removal of long-lived radionuclides from the biosphere by means of transmutation requires individual separation reprocessing for subsequent transmutation in reactors or accelerators;

- the extra durable matrices for immobilization of long-lived radionuclides for long-term safe storage and disposal require selective or fractional separation of the latter;

- the existing technologies for HLW management with vitrification and subsequent monitored storage of glass blocks require the separation of radionuclides from the bulk of non-radioactive waste in order to achieve higher glass quality, process safety, and cost effectiveness in the operations of glass boiling and storage of solidified materials.

2. R&D on Recovery of Radionuclides

Systematic studies and developments of efficient methods for separation of long-lived radionuclides from HLW from spent fuel reprocessing have been conducted in Russia for the last 20 years. These studies are carried out at several Russian Institutes (V.G. Khlopin Radium Institute, Institute of Chemical Technology, Institute of Physical Chemistry, Institute of Geochemistry and Analytical Chemistry, and other institutions) in collaboration with the radiochemical plants of MINATOM.

The key operations of the Russian technologies for separation of long-lived radionuclides involve selective recovery of cesium, strontium, technetium, and rare-earth and transplutonium elements, as well as residues of uranium, neptunium, and plutonium from the PUREX process. Different processes (precipitation, sorption, extraction, chromatography, etc.) for recovery of these components are under development in other parts of the world as well. This chapter considers those developments (first of all, extraction developments), which are now thought to be the most feasible ones for application.

2.1. Monodentate organophosphorus reagents for recovery of actinides and technetium

Efficient recovery of actinides and lanthanides from HLW may be achieved by means of monodentate neutral organophosphorus extractants. Among them, phosphine oxides possess the greatest extraction ability; due to high solubility of these compounds in organic diluents, preference is given to different-radical phosphine oxide (POR), i.e., isoamyl dialkylphosphine oxide wherein alkyl radicals have normal structure with 7-9 carbon atoms (Figure 5-1) [2,3].

KRI and INEEL jointly tested this extractant on HLW (Figures 5-2 and 5-3), and obtained a high recovery rate for actinides (up to 99.9%). Solid extractants, where the macroporous polystyrol-divinylbenzene matrix contains up to 50% of phosphine oxide, were synthesized on the basis of POR. Trials of this sorption variant were conducted in Russia on an industrial scale in a 60:1 column. The results confirmed efficient concentration of actinides in desorbate as well as high recovery rates. A distinguishing feature of the POR-based reactants is the possibility for recovery of technetium along with actinides. The recovery rate of technetium from actual HLW was over 90% [4]. These results were attained during the joint tests conducted by KRI and INEEL with INEEL HLW containing sodium bearing and dissolved calcines.

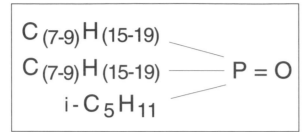

Figure 5-1.
Structure formula of POR.

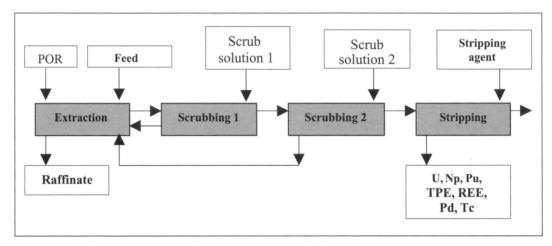

Figure 5-2. Flowsheet for HLW reprocessing by POR. Variant 1.

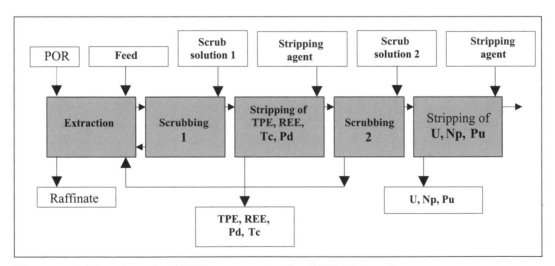

Figure 5-3. Flowsheet for HLW reprocessing by POR. Variant 2.

2.2. TRUEX process for recovery of actinides and lanthanides

The possibilities of the TRUEX process based on the use of bifunctional neutral organophosphorus compounds have been studied thoroughly. An interesting modification of the TRUEX process was developed by the Khlopin Radium Institute in collaboration with the Institute of Geochemistry and Analytical Chemistry [5]. In the classical TRUEX process, phenyloctyl-isobutyl-carbamoylphosphine oxide is diluted by dodecane with the addition of TBP as solubilizer. In the proposed modification, diphenyldibutyl-carbamoylphosphine oxide diluted by a polar organofluoric compound (fluoropole) is used as the extractant (Figure 5-4) [6]. Deletion of a solubilizator from the extractant composition eliminates the necessity for a washing operation to free the extractant from TBP destruction products. The technology of the modified TRUEX process was tested (Figure 5-5) to assess the possibility for efficient recovery of actinides and lanthanides [7]. The recovery rate for actinides and lanthanides was greater than 99.5% at a concentration factor of 4-6 and at purification coefficients for iron, zirconium, and molybdenum above 50. These results were confirmed by tests under static conditions, but not on actual industrial HLW [8].

PhOct/iBu$_2$. Phenyloctyl-N,N-diisobutylcarbamoyl phosphine oxide

Ph2/Bu$_2$. Diphenyl-N,N-dibutylcarbamoyl-phosphine oxide

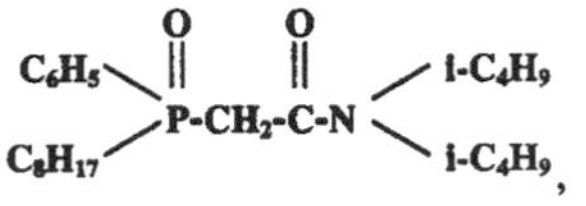

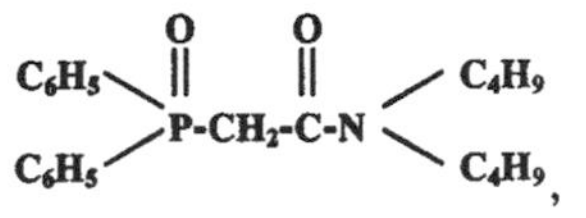

Figure 5-4. Classical and Russian options of carbamoyl-phosphine oxides used for TRUEX-process.

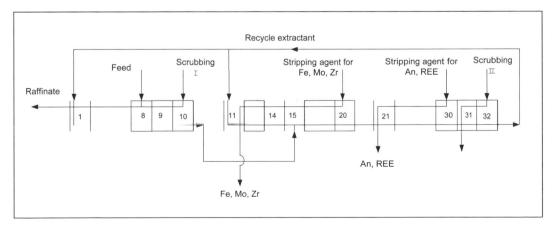

Figure 5-5. Flowsheet for An and REE recovery from UE-35 raffinate by the Russian TRUEX-process.

2.3. Chlorinated Cobalt Dicarbollide in HLW reprocessing

Use of chlorinated cobalt dicarbollide (ChCoDiC) in a polar diluent as an extractant is now the most widely promoted technology for recovery of radionuclides from HLW in Russia. Fundamentals of this extraction process were developed at the Khlopin Radium Institute and the Institute of Nuclear Research in the Czech Republic [9], whereupon KRI in collaboration with the Mayak facility brought this development to the level of industrial use at the radiochemical plant RT-1 [10].

ChCoDiC in a polar diluent efficiently extracts cesium from aqueous solutions; addition of polyethylene glycol (PEG) to the extractant makes it possible to extract strontium as well as TPE and rare-earth elements. The selective recovery of cesium was tested on INEEL HLW [2,3], resulting in a recovery rate of 99.998%. Different variants of the flow-sheets with combined and separate recovery of cesium, strontium, TPE and rare-earth elements were put through a series of tests from laboratory to pilot-industrial scale [9-12].

The ChCoDiC-based process for recovery of cesium, strontium, TPE, and rare-earth elements was tested in a hot cell test facility of KRI on 24 liters of actual raffinate from reprocessing of WWER-1000 spent fuel. The recovery rates were: Cs—more than 99.999%; Sr—99.998%, actinides—over 99.992%. The concentration degree of recovered radionuclides was equal to 6. The pilot-industrial tests of the ChCoDiC process at the Mayak facility confirmed the high efficiency of recovery of long-lived radionuclides from HLW. As a result of these tests, megacurie quantities of cesium and strontium were recovered, and a TPE concentrate containing 240 g of americium-241 and 21 g of curium-244 was obtained as well [12].

2.4. Industrial facility for HLW partitioning at the Mayak facility

The greatest achievement in the use of the ChCoDiC process in Russia is the introduction of a partitioning technology for HLW with different composition at the Mayak facility. By means of this technology, the first commercial facility in the world, UE-35, for recovery of radionuclides began operation in August 1996. The first line of this facility is designed for selective recovery of cesium and strontium from HLW.

Over 1000 m^3 of HLW have been reprocessed; the concentrates of cesium and strontium with a total activity of about 15 mln Ci were produced. The average recovery rates for cesium and strontium were 98.5%. The cesium and strontium concentrates were subsequently vitrified; this made it possible to increase the specific activity of the glass blocks up to 550 Ci/kg. Thus, an increase in cost for the partitioning facility by 5% led to a decrease of the production cost of high-level active glass by 60%.

For the second line of the UE-35 facility at the Mayak facility, which is intended for efficient recovery of actinides and technetium along with cesium and strontium from HLW, three feasible extraction processes are under investigation.

The first process uses POR to recover actinides, rare-earth elements, and technetium from raffinates arising from recovery of cesium and strontium by ChCoDiC. This process was successfully tested by joint efforts of specialists from the Khlopin Radium Institute and from INEEL at a setup of centrifugal contactors with the use of acidic INEEL HLW [2-4]. The second process for separation of actinides and rare-earth elements from HLW at Mayak is based on the Russian TRUEX process described above. The results of dynamic tests on simulated HLW from Mayak were confirmed by static tests on actual raffinates arising from the recovery of cesium and strontium at the UE-35 facility. In the course of these tests, 99.7% of alpha-nuclides were recovered. The third alternative for the second line of the UE-35 facility at Mayak envisions the use of a unified process for recovery of cesium, strontium, actinides, and rare-earth elements. The fundamentals of this process were elaborated by specialists of the Khlopin Radium Institute and INEEL and were brought to the level of dynamic tests in multi-stage centrifugal contactors on acidic HLW from INEEL [13,14]

2.5. Universal extractant for separation of cesium, strontium, actinides, and rare-earth elements from HLW (UNEX process)

The universal extractant (UE) is a mixture containing chlorinated cobalt dicarbollide, diphenyl-dibutyl-carbamoyl phosphine oxide with addition of polyethylene glycol, and a specific diluent (Figure 5-6). This solvent effectively extracts cesium, strontium, uranium, neptunium, plutonium, americium, curium, and rare-earth elements from acidic HLW.

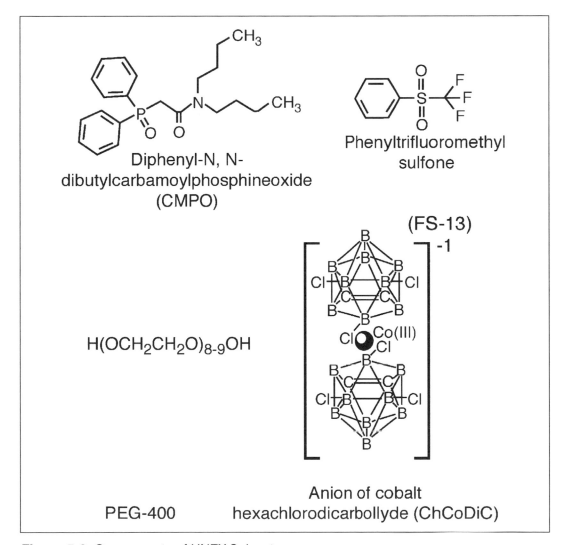

Figure 5-6. Components of UNEX-Solvent.

This process (UNEX) was tested on INEEL HLW using a 24-stage cascade of centrifugal contactors (Figure 5-7). Enough long-lived radionuclides were recovered for reclassification of the raffinate from HLW to LLW. Due to the possible introduction of the universal extractant at the Mayak facility in Russia and the Idaho National Laboratory in the U.S., efforts are underway to optimize this process. Problems of operational safety (i.e. explosion and fire safety as well as toxicity and corrosiveness) are also under study. These problems have already been elaborated in detail in relation to the use of ChCoDiC to recover cesium and strontium.

This example of a cooperative development at KRI and INEEL has demonstrated the efficiency and mutual benefits of cooperation between Russian and American specialists. The ChCoDiC process developed and implemented at the Mayak facility for Cs and Sr recovery from HLW was used as the basis of this cooperation. In the course of this cooperation, the UNEX process was developed as one of the variants for reprocessing INEEL HLW. After successful demonstration of the UNEX process with HLW at INEEL, this technology is being considered as an option for recovery of actinides and rare-earth elements in addition to Cs and Sr at the Mayak facility.

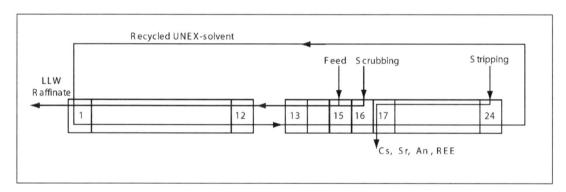

Figure 5-7. UNEX flowsheet for extraction recovery of long-lived radionuclides from HLW.

3. Conclusions

In order to reduce the volumes of HLW being vitrified, a facility for recovery of long-lived radionuclides has been constructed at the reprocessing plant of the Mayak facility. Efficient separation technologies are under development for reducing HLW volumes and for obtaining fractions of individual radionuclides. This creates the conditions needed for the future use of new methods for the management of the most hazardous radionuclides by transmutation or by final disposal in the form of very strong matrices.

Several Russian institutes are developing sorption and extraction processes that provide the efficient recovery of radionuclides. The most advanced method for HLW reprocessing is extraction by chlorinated cobalt dicarbollide. This method, which affords efficient recovery of cesium and strontium, has been introduced on an industrial scale at Mayak (UE-35 facility). The introduction of the first partitioning stage has already resulted in economic savings from vitrification of reduced volumes of HLW.

Three process variants for the second line of the UE-35 facility are now under development for the recovery of actinides, rare-earth elements, and technetium. The first variant is based on POR, the second on the Russian modification of the TRUEX process, and the third on the use of the universal extractant (mixture of ChCoDiC, CMPO, and PEG in special diluent). This UNEX process has been developed through KRI and INEEL cooperation and has been investigated using tests under dynamic conditions with INEEL HLW. Russian specialists are currently working on separation technologies to obtain individual fractions of long-lived radionuclides to assure the safety of processes and equipment being used.

References

1. Romanovsky, V.N. (1994) *Proceedings of a Regular Advisory Group Meeting* (31 August – 3 September, 1993, IAEA – TECDOC-732), 75.
2. Todd, T.A., Brewer, K.N., Herbst, R.S., Law, J.D., Romanovsky, V.N., Lazarev, L.N., Smirnov, I.V., Esimantovskiy, V.M., and Zaitsev, B.N. (1995) In *Proceedings of the Fifth International Conference on Radioactive Waste Management and Environmental Remediation* (ICEM'95) vol. 1, Berlin, Germany, 463.
3. Todd, T.A., Brewer, K.N., Herbst, R.S. Law, J.D., Romanovsky, V.N., Lazarev, L.N., Smirnov, I.V., Esimantovskiy, V.M., and Zaitsev, B.N. (1996) In *Value Adding through Solvent Extraction: Proceedings of ISEC'96, International Solvent Extraction Conference* (March 17-21, 1996), vol. 2., University of Melbourne, Melbourne, Australia, 1303.
4. Law, J.D., Herbst, R.S. Todd, T.A. Brewer, Romanovsky, V.N., Esimantovskiy, V.M., Smirnov, I.V., Babain, V.A., Zaitsev, B.N., and Dzekun, E.G. (1996) In *Proceedings of the International Topical Meeting Nuclear Hazardous Waste Management: SPECTRUM '96.* American Nuclear Society, La Grange Park, IL, 2308.
5. Horwitz, E. P., and Schulz, W. W. (1998) Solvent Extraction in the Treatment of Acidic High-Level Liquid Waste: Where Do We Stand? In *Metal Ion Separation and Preconcentration: Progress and Opportunities.* Bond, A.H., Dietz, M.L., Rogers, R.D. (eds.) American Chemical Society.
6. Romanovsky, V.N., Shadrin, A., Smirnov, I.V., Babain, V., Pribulova, G., Litvina, M. (1994) *SPECTRUM'94 Proceedings* (August 14-18, 1994, Atlanta, Georgia, U.S.A.), vol. 2, 833.
7. Myasoedov, B.F., Chmutova, M.K., Smirnov, I.V., and Shadrin, A.U. (1993) In *Global '93: Future Nuclear Systems: Emerging Fuel Cycles and Waste Disposal Options.* American Nuclear Society, La Grange Park, IL, 581.
8. Romanovsky, V.N., Smirnov, I.V., Shadrin, A.Y., Babain, V.A., Murzin, A.A., Myasoedov, B.F., Chmutova, M.K., Logunov, M.V., Mezentcev, V.A., and Polun, A.K. (1998) In *Proceedings of the International Topical Meeting on Nuclear Hazardous Waste Management: SPECTRUM'98.* American Nuclear Society, La Grange Park, IL, 576.
9. Galkin, B.Ya., Esimantovskiy, V.M., Lazarev, L.N., Lyubtsev, R.I., Romanovsky, V.N., Kyrs, M., Rais, J. et al. (1988) *ISEC-88* vol. 4. Moscow, Nauka, 215.
10. Dzekun, E.G., Glagolenko, Y.V., Drojko, E.G., Kurochkin, A.I., Scobtsov, A.S., Domnin, V.V., Logunov, M.V., Rovnyi, S.I., Suslov, A.P., Romanovsky, V.N., Esimantovskii, V.M., Lazarev, L.N., Lyubtsev, R.I., Filippov, E.A., Nardova, A.K., and Mamakin, I.V. (1996) In *Proceedings of the International Topical Meeting Nuclear Hazardous Waste Management, SPECTRUM'96.* American Nuclear Society, La Grange Park, IL., 2138.
11. Low, J.D., Herbst, R.S., Todd, T.A., Brewer, K.N., Romanovsky, V.N., Lazarev, L.N., Smirnov, I.V., Babain, V.A., Zaitsev, B.N., and Dzekun, E.G. (1995) *Flowsheet Development Studies Using Cobalt Dicarbollide and Phosphine Oxide Solvent Extraction Technologies for the Partitioning of Radionuclides from ICPP Sodium-bearing Waste.* Idaho National Engineering Laboratory, Idaho Falls, Idaho.
12. Esimantovskiy, V.M., Galkin, B.Ya., Dzekun, E.G., Lazarev, L.N., Lyubtsev, R.I., Romanovsky, V.N., and Shichkin, D.N. (1992) *Proceedings of the Symposium on Waste Management* vol. 1 (Tucson, Arizona, March 1-5, 1992), 805.

13. Todd, T.A., Brewer, K.N., Law, J.D., Wood, D.J., Herbst, R.S., Romanovsky, V.N., Esiman-tovskiy, V.M., Smirnov, I.V., and Babain, V.A. (1998) *Proceedings of 13th Radiochemical Con-ference* (Marianske Lasne, Czech Rep).
14. Todd, T., Brewer, K., Law, J., Herbst, R., Romanovsky, V., Esimantovskiy, V., Smirnov I., Babain, V., and Zaitsev, B. (1998) In *Proceedings of the International Topical Meeting on Nuclear Hazardous Waste Management: SPECTRUM'98* (Denver, Colorado), 743.

Valery N. Romanovsky is Deputy Director General of the Khlopin Radium Institute in St. Petersburg where he has been working since 1969. His field of expertise is separation processes and technologies in HLW management and in spent nuclear fuel reprocessing. He received his Ph.D. in Chemistry from the Technical University in St. Petersburg in 1970. He is a member of the Joint Committee for Environmental Restoration and Waste Management, which coordinates cooperative R&D between the U.S.A. and Russia in this field.

CHAPTER **VI**

RADIATION PROTECTION ASPECTS OF NUCLEAR WASTE SEPARATIONS

DONALD T. OAKLEY
Institute for International Cooperative Environmental Research
Florida State University
Tallahassee, Florida 32310, U.S.A.

ABSTRACT

This paper provides an overview of radiological health impacts of nuclear waste separations and inter-related disposal considerations. Three principal areas are covered: 1) comparison of radiological impacts of nuclear fuel cycle options, including separations, 2) radiation protection issues related to separations and resulting waste, and 3) management of long-lived radionuclides. The near-term risks are related to occupational and public radiation exposures during the operation of separations facilities. Because long-lived radionuclides are generated in uses of nuclear energy, the longer-term risks are due to the lack of safe isolation of nuclear wastes in the environment.

1. Occupational and Public Radiation Exposure from Two Nuclear Fuel Cycle Options

The Nuclear Energy Agency (NEA) of the Organization for Economic Cooperation and Development has recently completed a report on the radiological impacts of spent fuel management options [1]. This report addresses the complicated technological, health, and societal issues associated with nuclear waste by-products of the nuclear fuel cycle. This work evaluates technology options and their use under current radiological control practices, and then projects life-cycle impacts at present and into the near future. Radiological impacts were examined for two spent fuel management options: 1) reprocessing of spent fuel and one-time recycling of separated plutonium in the form of mixed oxide fuel (MOX), and 2) a once-through fuel cycle in which the spent fuel is not reprocessed but is considered as waste. Numerous assumptions were made concerning fuel burnup, waste handling, and geologic disposal for HLW and spent fuel. The starting point for each option was a 1000 MW(e) pressurized water reactor (PWR) with approximately 40 GWd per ton of fissionable material burnup fuel.

In this NEA 2000 study, radiological impacts are compared within the framework of the recommendations of the International Commission on Radiological Protection (ICRP) as described in its Publication 60 [2]. Radiological impacts have been evaluated for members of the general public and for workers and are presented in terms of critical group doses and collective doses arising from the various stages of the two fuel cycles considered. Collective doses are for regional populations, which, in all cases except for mining and milling, correspond to the population of Europe.

Collective doses to the public are summed over 500 years and are normalized to the electricity produced in gigawatt-years. There is an ongoing debate within the scientific community regarding the use of collective doses, particularly in the case of small individual doses received by large populations and over many generations into the future [3]. A clear consensus has not emerged on this important issue. Given that there is no consensus on this point, the NEA expert group decided to proceed along ICRP Publication 60 [2] lines. The effect of this assumption on long-term considerations will be discussed later.

In calculating doses to members of the public, the NEA 2000 report makes assumptions about population distribution, habits of individuals, characteristics of the environment in which they live, and conditions of releases. These assumptions can have a considerable influence on the magnitude of calculated doses. Site-specific calculations were based on actual operating experience at European facilities.

Occupational doses received over the entire fuel cycle are dominated by doses to workers at the nuclear power plant. Separations required for the second option have a negligible impact on worker exposure. The occupational doses to workers in nuclear power plants are not affected by the type of fuel used (UO_2 or MOX). At the fuel fabrication stage, there is a significant difference between occupational exposures of the two fuel cycle options. However, the absolute values at that stage are only a small fraction of the sum over the whole fuel cycle for both options.

In NEA 2000, impacts on the public are calculated by dividing the fuel cycle into four stages: uranium mining and milling, fuel fabrication (including enrichment and uranium conversion), power production, and reprocessing. The assessed collective doses for mining and milling and for reprocessing are similar for both the impacts on the general public and on fuel cycle facility workers. Results of this study show that the highest radiological impacts come from uranium mining/milling and reprocessing. Individual doses to members of the critical group from power production are much lower than those from mining and milling or from reprocessing. Fuel fabrication gives rise to the lowest collective doses of any stage

A summary of all occupational and public radiological impacts, from each step of both fuel cycles, is presented in Table 6-1 (NEA 2000). The mining and milling, power production, and reprocessing stages dominate the collective doses to the public. While power production causes the same radiological impacts for both fuel cycle options, the variations in the radiological impacts of the other two stages tend to be in opposite directions and offset each other. Through reprocessing and through the use of MOX-fuel, the need for natural uranium could be reduced by about 21%, and consequently the public and worker exposures caused by the mining and milling stage could be reduced in the same proportion. On the other hand, the reprocessing stage adds to the public and worker collective doses.

Public collective doses estimated in NEA 2000 and the overall radiological comparison of options are seen to be highly sensitive to the assumptions regarding good mill-tailing pile management. The collective dose summed over 500 years to the regional population

TABLE 6-1. Public and Occupational Radiation Exposure [1]

Fuel cycle stage	Public (calculated) Collective dose truncated at 500 years (manSv/Gwa)		Workers (operational data) Annual collective dose (manSv/Gwa)	
	Once-through	Reprocessing	Once-through	Reprocessing
Mining and milling[1]	1.0 (1-1000)	0.8 [0.8x1-1000)]	0.02-0.18	0.016-0.14
Fuel conversion and enrichment	0.0009		0.008-0.02	0.006-0.016
Fuel fabrication			0.007	0.094
Power generation	0.6	0.6	1.0-2.7	1.0-2.7
Reprocessing, vitrification	Not applicable	1.2	Not applicable	0.014
Transportation	Trivial	Trivial	0.005-0.02	0.005-0.03
Disposal	0	0	Trivial	Trivial
Total	**1.6**	**2.6**	**1.04 – 2.93**	**1.14 – 2.99**

[1] Estimates of public exposure accumulated over 500 years can vary significantly depending on modeling assumptions regarding tailings-pile maintenance and population characteristics.

(i.e., within a radius of 2000 km) is up to around 1 manSv/Gwa (mansievert/gigawatt-year) for uranium mining and milling and a maximum of about 1.2 manSv/Gwa for reprocessing. Available site-specific calculations support the conclusions of the generic calculations in the sense that the assessed collective dose of an actual reprocessing facility is 0.6 manSv/Gwa, and that the critical group doses are higher than for the other two stages (i.e., power plant operation and fuel fabrication) of the fuel cycle that were considered.

For both the uranium mining and milling and the reprocessing stages, the assessed critical group doses are in the range of 0.30 to 0.50 mSv/a (millisievert/year). Actual critical group doses at specific sites can be significantly different, due to differences in the habits and location of local populations. Overall, the differences between the two fuel cycles examined in the report are small from the standpoint of radiological impact on humans. For purposes of comparison, the average natural radiation background dose to humans is 2.4mSv/a. This corresponds to a collective dose for 1000 people of 2.4 manSv/a.

2. Uncertainties in Calculations of Health Impacts of Nuclear Fuel Cycle Operations

2.1. Environmental conditions

Large uncertainties are associated with public-exposure estimates because of assumptions regarding models, scenarios, and input values. In particular, uranium mining is very site-specific and doses are strongly influenced by environmental conditions such as characteristics of the uranium-bearing rock, mining and milling practices, long-term stability of disposed tailings, and procedures for maintenance and remedial actions. Actual ^{222}Rn emanation rates could be significantly different from those assumed in this study, leading to higher or lower collective doses. Indeed, if the tailing piles were partially uncovered following a period of poor maintenance, collective doses of up to a few tens of mansieverts per gigawatt-year would be possible.

2.2. Time period for dose estimation

Collective doses to members of the public require critical examination. For example, in the NEA 2000 study collective doses have been calculated to 500 years into the future. These doses may be calculated for longer time periods, including infinity, but considerable controversy exists over the practicality of doing this. As collective doses from major components of the fuel cycles (i.e., mining and milling, power production, and reprocessing) involve long-lived radionuclides, such an approach would lead to larger collective doses. For example, the majority of the collective doses calculated for the assumed 500-year period for both power production and reprocessing arises from the relatively long-lived, mobile radionuclide ^{14}C (half-life 5,730 years). This isotope will continue to contribute to collective doses long after the assumed calculation period.

For the mining and milling stage, the production of ^{222}Rn in tailings will continue at a slowly declining rate for a period of hundreds of thousands of years, because it is generated by the decay of the long-lived radionuclide ^{230}Th (half-life 77,000 years). The contribution to the collective doses from this part of the fuel cycle would be significant if calculated for very long time periods >500 years. Safe maintenance of tailings from uranium mining and milling for very long periods has yet to be demonstrated.

Disposal of HLW and spent fuel in geologic repositories will result in calculated collective doses that do not begin to accumulate until well after 500 years. The principal long-term dose contributors are ^{126}Sn, ^{226}Ra, ^{79}Se, ^{129}I, ^{135}Cs, ^{99}Tc, and ^{237}Np and other long-lived nuclides. Improvements in separation technologies and alternative disposal means for very long-lived isotopes may provide a crucial step in the acceptance of geologic disposal. Under some assumed repository conditions, examples of radionuclides that would begin to contribute significantly to the collective dose tens of thousands of years following disposal are shown in Table 6-2 (NEA 2000).

2.3. Non-radiological health impacts

Evaluations of nuclear fuel cycle impacts frequently concentrate on radiological impacts, which may not be as significant as impacts from competing technologies or exposure to other unsafe materials or conditions. For example, mining and milling activities historically have had one of the higher industrial accident rates in the United States.

Decontamination and decommissioning of facilities closely resemble construction work, another high-risk occupation. These non-radiological effects should be considered when comparing relative risks of technological options. Further, to add balance to additional trade-offs in technology evaluation, considering the impacts of technologies that are alternatives to the nuclear fuel cycle would be useful.

3. Other Approaches

One of the principal obstacles facing disposal options for long-lived radionuclides is the difficulty of projecting human impacts far into the future. A recent analysis by the U.S.

TABLE 6-2. Separation and Transmutation of Key Radionuclides [5]

Radio-nuclide	Environmental impact	Partitioning technology	Transmutation technology
^{126}Sn	Important in groundwater release pathway.	Probably goes mainly to dissolver insolubles in reprocessing. Difficult to separate from	Half time of transmutation is very long even in high neutron fluxes.
^{79}Se	Important in groundwater release pathway.	Probably difficult to separate from HLW.	Long transmutation half time for ^{79}Se. More ^{79}Se would be formed from ^{78}Se.
^{135}Cs	Important in groundwater release pathway.	Methods for separation of cesium from HLW have been developed.	Not feasible because of transmutation of other cesium isotopes, forming more ^{135}Cs.
^{241}Am	Important in groundwater release pathway as parent of ^{233}U.	Much research in progress on separation from HLW by aqueous methods. Separation from curium and lanthanides requires further development.	Could be incinerated, preferably in fast reactors. Multiple cycling & intermediate reprocessing needed. Curium would be generated.
^{237}Np	Important in groundwater release pathway as parent of ^{233}U.	Probably could be separated during conventional reprocessing.	Could be incinerated, but significant amounts but significant amounts of ^{238}Pu would be formed.
^{99}Tc	Important in human intrusion and groundwater release scenarios.	Divided between dissolver insolubles and dissolver liquor. Could be separated from the latter during reprocessing.	Could be transmuted. Several cycles with intermediate processing needed.
^{129}I	Important in groundwater release pathway.	Mainly diverted into LAW streams in reprocessing. Could be largely separated if required.	Could be transmuted. Several cycles with intermediate processing probably needed. Target stability could be a problem.

National Research Council of long-term management of U.S. DOE waste sites [4] observes that ". . . there is no convincing evidence that institutional controls and other stewardship measures are reliable over the long term." The report further notes that "Other things being equal, contaminant reduction is preferred to contaminant isolation and the imposition of stewardship measures whose risk of failure is high."

Concerns over safe management of long-lived radionuclides have lead to a recurring interest, primarily in the scientific community, in evaluation of alternatives for waste disposal. Major improvements in aqueous and non-aqueous separations technologies would be required for these schemes to progress. Two examples follow of contaminant reduction and isolation.

One of these two schemes, nuclear waste transmutation, has been the subject of recent comprehensive reviews (see Chapter VII). This process is postulated to reduce contaminants by transmuting long-lived nuclides into stable or less radioactive elements. The Nuclear Energy Agency has recently completed surveys [5, 6] of worldwide progress in this area. In addition, the U.S. Department of Energy has prepared a plan for developing transmutation [7] that provides an overview of reactor and accelerator-based options for transmutation of nuclear waste. Taken together, these analyses summarize the significant technology advances and long transmutation periods required for transmutation to succeed. Although complete burnup of nuclear wastes cannot be achieved even with transmutation, long-term dose consequences of geologic waste disposal are projected to be reduced by as much as ~90% compared to non-transmuted waste. To achieve this level of reduction would require significant improvements in partitioning of long-lived radionuclides and the ability to fabricate stable targets for neutron irradiation. Debates continue over the viability of the concept as an integrated means to produce power while treating waste.

Another scheme involves the separation of long-lived nuclides and sending them into outer space. This option was analyzed in the 1970s, when it was studied as a waste disposal option. Again, this technique would depend on significant improvements in waste separations. It would also require vastly increased confidence in the technology to launch waste-bearing vehicles into outer space and to survive accidents during launch. Both options are expensive, and significant technology breakthroughs are necessary for either to succeed. However, within 50 to 100 years major advances are not out of the question for these and other methods. Additional study of these and other options may be warranted by the high cost of long-lived radionuclide disposal and public acceptance of long-term safety assurances. Means may evolve to reverse biological damage due to radiation effects, thus reducing the risk of the presence of long-lived radionuclides in the environment.

4. Conclusions

The radiological impact of current nuclear waste separations contributes moderately to the overall impact of the entire nuclear fuel cycle, not counting radiological impacts of long-term HLW disposal. The overall near-term fuel cycle impacts to the public are relatively minor. Little difference exists in the radiological impact of reprocessing nuclear fuel *vs.* once-through fuel use. Longer-term impacts are projected to be moderate as well, although questions continue to be raised about institutional and scientific aspects of waste isolation over long time periods.

Proposed improvements in waste partitioning, such as non-aqueous means for separations, hold promise for significantly reducing the volume of secondary wastes that must be either treated or isolated to prevent public radiation exposure. Improved separation methods will also be needed to support new technologies that have been proposed and others that might be proposed to reduce long-term concerns for waste isolation. As advances are made towards smaller, robotically controlled separation processes, both short- and long-term occupational exposures can be expected to decline.

References

1. NEA (2000) *Radiological Impacts of Spent Nuclear Fuel Management Options*. Nuclear Energy Agency, Paris, France.
2. ICRP (1991) *1990 Recommendations of the International Commission on Radiological Protection*. ICRP Publication 60, Pergammon Press.
3. ICRP (1998) *Radiological Protection Policy for the Disposal of Radioactive Waste*. ICRP Publication 77, Annals of the ICRP, vol. 27 Supplement, Pergammon Press.
4. NRC (2000) *Long-Term Institutional Management of U.S. Department of Energy Legacy Waste Sites*. National Academy Press.
5. NEA (1999a) *Actinide and Fission Product Partitioning and Transmutation*, Proceedings of the Fifth International Information Exchange Meeting, November 25-27, 1998, Nuclear Energy Agency, Paris, France.
6. NEA (1999b) *Actinide and Fission Product Partitioning and Transmutation, Status and Assessment Report*. Nuclear Energy Agency, Paris, France.
7. DOE (1999) *A Roadmap for Developing Accelerator Transmutation of Waste ATW) Technology*. U. S. Department of Energy, DOE/RW-0519.

Donald T. Oakley has over 35 years of experience in environmental research and waste management with the U.S. Public Health Service, the U.S. Environmental Protection Agency, and the Los Alamos National Laboratory. He is currently an Associate Director of the Institute for International Cooperative Environmental Research, Florida State University. His research activities have focused on environmental effects of nuclear weapons testing, nuclear power applications in space, and high-level nuclear waste disposal. His current research interests include environmental technology development and deployment, and related international environmental applications. He is a member of the High-Level Waste Tanks Advisory Panel, which evaluates the safety of waste storage and treatment at the U.S. Department of Energy's Hanford Site. Dr. Oakley received his Sc.D. in Environmental Health Sciences from the School of Public Health, Harvard University. He is a Registered Professional Engineer and a Certified Health Physicist.

CHAPTER **VII**

ACCELERATOR-DRIVEN TRANSMUTATION TECHNOLOGIES FOR NUCLEAR WASTE TREATMENT

HANS S. PLENDL
Department of Physics
Florida State University
Tallahassee, Florida 32306, U.S.A.

ABSTRACT

Some of the disadvantages of geological disposition of nuclear waste materials can be overcome by transmutation of the longest-lived and environmentally most dangerous nuclides into shorter-lived and less toxic ones. This is already being done in some reactors where mixtures containing plutonium isotopes are used as fuel. Transmutation of nuclear waste materials in accelerator-driven subcritical reactor systems would have a number of advantages since it would use accelerator-produced neutrons from the spallation process rather than reactor-produced neutrons from the fission process.

This technology has been studied for decades at many laboratories. Until the early 90s, it was not a serious contender for applications because of what appeared to be inherent technical obstacles. During the 90s, however, the need for accelerators producing up to 100 times more powerful proton beams than the most powerful ones in existence and for accurate cross sections at energies up to several orders of magnitude higher than the conventional reactor energy range were being met. Accelerator-driven transmutation technologies also began to gain the support that will be needed to overcome the remaining obstacles as it was gradually realized that geological depositories need to be augmented by other methods of nuclear waste disposition if they are to be practical in the long run.

Paramount among the problems that still await a solution is the need for efficient methods of partitioning and separating those radionuclide species that must be re-circulated after each of many necessary irradiations from the shorter-lived ones that can be disposed of at that stage. This problem is far more severe for accelerator-driven reactor systems than for conventional ones partly because the reactions in such systems occur at much higher particle energies and result, therefore, in dozens of radionuclides and a myriad of chemical compounds that are not produced at the lower operation energies of conventional reactors. Recently developed non-aqueous separation methods may offer a solution to this vexing problem, which has caused some scientists, and nuclear chemists in particular, to be highly skeptical of accelerator-driven transmutation technologies as an option to complement other methods of nuclear waste disposition.

1. Accelerator-driven Systems vs. Conventional Reactors

Accelerator-driven systems (ADS) for transmutation of radioisotopes and/or production of thermal energy are basically reactors that are run in a subcritical mode. They use neutrons that are produced in spallation processes induced by proton beams from accelerators in suitable target materials, rather than neutrons that are produced in fission processes in conventional reactors. While conventional reactor operation can be sustained only in a critical mode, accelerator-driven transmutation systems can be run in a subcritical mode because the neutrons needed to sustain the operation are supplied from the outside. Hence accelerator-driven transmutation systems are intrinsically safer to operate.

Some of the other advantages of the spallation process and of the accelerator-driven systems employing it are:

- More neutrons are produced at less energy to be dissipated.

- Much less radioactive waste and excess fissionable material is produced, in particular the production of Pu and other fissionable actinide isotopes can be avoided so that the proliferation problem is minimal.

- For energy production, low-quality fissionable materials such as thorium can be used.

- Closed cycle operation is feasible.

- Such systems are more flexible than reactors, in particular in regard to the fuel cycle employed in a system intended for transmutation/energy production.

Among the apparently inherent obstacles to constructing systems employing the spallation process for nuclear waste transmutation or energy production on a practical scale are the needs for

- Accelerators that can provide proton beams of the order of 1 GeV and up to 100 mA, i.e., systems that can produce and dissipate a power output of up to 100 times that of the LANSCE facility (formerly known as LAMPF) at Los Alamos, the most powerful existing accelerator;

- Sufficiently accurate cross sections for specific nuclides and other nuclear data in the relatively unexplored intermediate energy range for macroscopic transport and activation calculations;

- Partitioning/separation methods that work for the short storage times available between the many recycling processes needed to transmute appreciable amounts of material;

- Lowering the cost of such a system;

- Overcoming the skepticism of the reactor and nuclear chemistry communities toward an untried new method when time-honored methods that are known to work are available.

Some of these problems can be overcome at the present time: accelerators of the necessary power have been designed, not only for waste transmutation or energy production [1], and can be built whenever the necessary funds become available. The needed nuclear data are being accumulated by experimentalists working at most of the intermediate energy installations in active use; and theorists and computer specialists have developed models and codes to calculate data that are difficult or impossible to obtain experimentally, so that extensive data libraries have been compiled and are being continually augmented. For detailed accounts of some of these efforts, see Refs. [2,3]. The other problems listed above will be considered in sections 3 and 4.

Accelerator-driven systems have been designed for a variety of purposes besides transmutation and thermal energy production. Some of these systems are being operated successfully [4] while design work on others had to be terminated because the project lost its funding [1]. Table 7-1 provides an overview and a comparison with some of the uses of conventional reactor systems.

TABLE 7-1. Uses of Accelerator-Driven Systems and Conventional Reactors

	Production primarily of:					Transmutation of:		
	Energy	Neutron Beams	Needed Radio-isotopes	Tritium	Pu	Unwanted Radio-isotopes	M.A.	Pu
Reactor:	✓	✓	✓	✓	✓	—	✓	✓
ADS:	✓	✓	✓	(✓)	—	✓	✓	✓
e.g:	EA	SNS		(APT)		ATW	ATW	ATW

M.A. - Minor Actinides -Np, Am, Cm
EA - Energy Amplifier [5] (CERN)
SNS - Spallation Neutron Source [4] (PSI, ORNL, EU, JAERI, and others)
APT - Accelerator Production of Tritium [1] (LANL/Savannah River; terminated)
ATW - Accelerator-Driven Transmutation of Nuclear Waste [2,3] (LANL, JAERI, and many other programs)

2. Examples of Current ADS Projects for Nuclear Waste Transmutation and Energy Production

The concept of an accelerator-driven system for nuclear waste transmutation/production of thermal energy is illustrated in Figure 7-1 [6]. The accelerator beam produces neutrons in the target material by spallation. These neutrons interact with the waste material in the blanket surrounding the target, transmuting some of it by fission and other processes. The waste material is continually recycled and reprocessed by separation and partitioning methods that need to work during relatively short time periods because a large number of recycling processes are needed to transmute any appreciable fractions of it. The separated material that cannot or does not have to be recycled is then prepared for geological disposal. As indicated in Figure 7-1, the thermal energy produced by the nuclear reactions is converted to electric energy, some of which is used to run the system and the rest of which is available for external use.

Four of the current major ADS R&D projects are compared schematically in Figure 7-2 [7]. Transmutation of long-lived isotopes of Pu, of the minor actinides, and of I and Tc isotopes from the waste of reactors that produce energy for civilian use is emphasized in the U.S., French, and Japanese projects; production of thermal energy is considered a useful by-product. In the CERN project, production of thermal energy in a safe way from

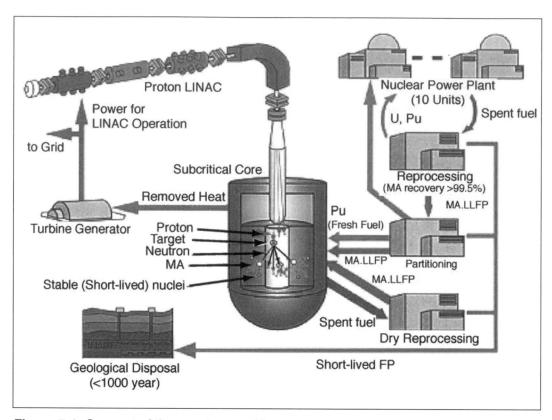

Figure 7-1. Concept of the accelerator-driven transmutation system [6].

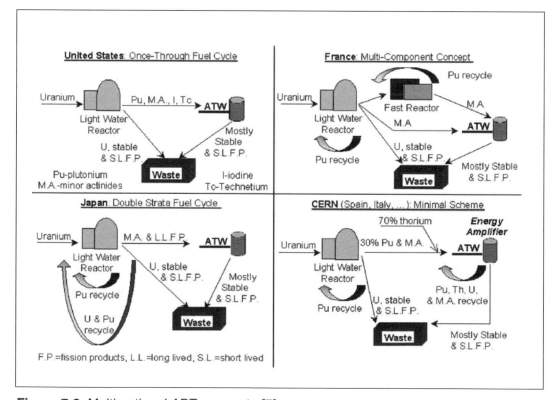

Figure 7-2. Multi-national ADT concepts [7].

plentiful fissionable material (Th) is emphasized, but reactor waste products can be transmuted at the same time. These and nearly all other currently active ADS projects are multinational and interdisciplinary in character.

Other R&D projects are being pursued in the United States and Russia and in nearly all countries having nuclear waste disposal problems, most of them in close cooperation with other programs [2,3]. They are supported by international organizations such as the EU and the ISTC, by national or regional governments, by private companies, or by a combination of the above. Some of these other projects are exploring all aspects of ADS for waste transmutation/ energy production, e.g., the recently initiated MYRRHA program at the national research center in Mol, Belgium. Others accomplish essential tasks by investigating specific problems such as comparisons of different types of spallation targets, subcritical assemblies, or cooling agents, e.g., the programs at research institutes in Rez (Czech Republic), Obninsk, Moscow, and St. Petersburg (Russia), and in Minsk (Belarus). Conceptual and computational studies are underway at a number of additional institutions, e.g., at the Technical University in Budapest, the Netherlands Energy Research Foundation laboratory in Petten, and at several U.S. national laboratories besides Los Alamos.

3. A Look to the Future

An extrapolation of the current R&D situation is shown in Figure 7-3 [7], a graphic representation of one of several scenarios developed as part of the Accelerator Transmutation of Waste (ATW) Program of the U.S. Department of Energy [8] for some of the long-lived portions of the waste materials from civilian nuclear power plants. Under the conditions

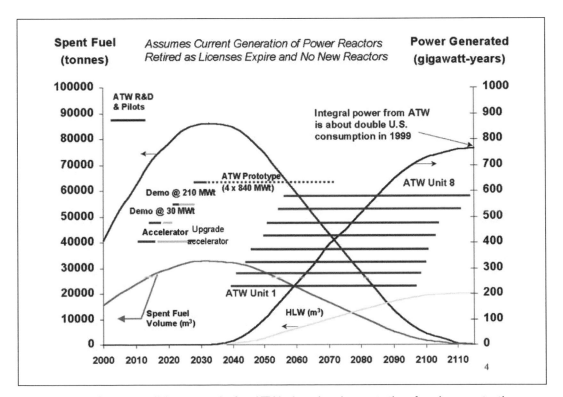

Figure 7-3. One possible scenario for ATW plant implementation for demonstration and production [7,8]

indicated in this figure, a small-scale demonstration unit would go on-line in 2012 to be followed by a larger one in 2020 that would eventually be converted into a prototype unit. As indicated in the figure, the first full-scale unit of this type would go on-line in 2038 and would be followed by seven additional ones. The total high-level waste produced by this system during 60 years of operation would be less than half (in terms of geological storage space needed) the total amount of the much more toxic spent fuel that would have to be stored in the absence of such a system. Once in full operation by 2060, the system would generate about 5% of the total electric power currently consumed annually in the U.S. and thus recoup some of its building and operating costs [7,8].

Scenarios of constructing and putting into operation assemblies of accelerator-driven systems for transmutation and energy production are being considered to meet national needs in other countries. They would require less investment and are likely to be put into practice sooner than the scenarios envisaged to meet the need of the United States for converting spent fuel from its civilian nuclear power plants into smaller and less dangerous amounts of high-level waste. The disposition of the huge Cold War stockpiles of surplus plutonium in the U.S. and in Russia by transmutation, however, may take longer to accomplish than the transmutation of spent fuel. But it may turn out to be less expensive to transmute these fissionable materials, in spite of the cost of transporting and safeguarding such materials, because larger amounts of thermal energy would become available than in the transmutation of non-fissionable materials.

4. Can Such a Future be Realized?

The most serious obstacle that must be overcome before any of these scenarios can be put into practice is the current lack of suitable partitioning/separation methods. In the initial stages of the development of ADS for transmutation/energy production, most nuclear physicists involved in that stage left the task of meeting this need to the nuclear chemistry community, which quickly realized that the existing separation methods were totally inadequate to accomplish such a task. A non-aqueous separation method like the one recently developed at Argonne National Laboratory (ANL) may eventually meet this need, but the present ANL method as outlined in Chapter IV would have to be considerably modified to meet the special requirements posed by the ADS transmutation process.

It may indeed be possible in the future to produce energy economically and on a large scale from what is now considered to be nuclear waste or surplus material. But much work will have to be done before such a future can become reality. Moreover, society will need to be convinced that the resulting high costs are necessary, since such efforts are bound to be even more expensive than the Manhattan Project for nuclear weapons production in the U.S. during World War II or the equivalent project in the former USSR. And such efforts would have to be supported for the several decades that may be needed to realize such projects rather than the few years it took to develop and build the first nuclear bombs. Thus, to deal with the legacy of waste that accumulated during the succeeding decades when the material for tens of thousands of such weapons was produced and reactors were run for civilian applications of nuclear energy, society and its leaders must be convinced to support projects that do not appear as necessary as the projects producing weapons of unprecedented power were to leaders of society at an earlier time. As scientists we know from experience how necessary the safe disposal of nuclear waste and the cleaner production of energy are. It is up to us to make that urgent need more generally known.

References

1. See, e.g., Lisowski, P. (2000) Accelerator Production of Tritium Program Status, *Proceedings of the Fourth Topical Meeting on Nuclear Applications of Accelerator Technology*, Washington. D.C., Nov. 13-15, 2000, in press;
   Lagniel, J.-M. (2000) Accelerateurs de haute intensité. *Ann. Phys. Fr.* **25**, 293.
2. Plendl , H.S. (ed.) (1998) Special Issue on Nuclear Transmutation Methods and Technologies for the Disposition of Long-Lived Radioactive Materials, *Nuclear Instruments and Methods in Physics Research* A, **414**, 1-126.
3. Plendl, H.S. (ed.) (2001) Special Issue on Accelerator-Driven Systems, *Nuclear Instruments and Methods in Physics Research* A, **463**, 425-666.
4. The spallation neutron source SINQ at the Paul Scherrer Institute (PSI) in Switzerland is an outstanding example; see the article by G. Bauer in Ref. [3].
5. Rubbia, C., et al. (1995) *Conceptual Design of a Fast Neutron Operated High Power Energy Amplifier*, CERN Report AT/95-44;
   Rubbia, C. (2000) Accelerator-Driven Sub-Critical Reactors for Radioactive Waste Transmutation and Energy Production, OECD Global Science Forum, *Proceedings of Workshop on Strategic Policy Issues Concerning the Future of High-Intensity Proton Beam Facilities*, Paris, 25-26 Sep. 2000, in press.
6. Sasa, T. personal communication.
7. Beller, D. et al., article in Ref. [3].
8. *A Roadmap for Developing Accelerator Transmutation of Waste (ATW) Technology*, U.S. DOE Report DOE/RW-0519 (1999).
   See also ATW Roadmap web site (http://www.pnl.gov/atw/).

Hans S. Plendl has over 40 years of experience in basic and applied nuclear physics at university and government laboratories in the United States, Switzerland, and Germany. Dr. Plendl received his Ph.D. in experimental nuclear physics from Yale University and his B.A. degree from Harvard University. He is Emeritus Professor of Physics at Florida State University. His current research activities are focused on the disposition of long-lived nuclear waste materials. As Senior Research Fellow at the Institute for International Cooperative Environmental Research, he provides technical support for the joint environmental restoration and nuclear waste management activities of the U.S. Department of Energy and the Russian Ministry of Atomic Energy (JCCEM Program).

Conclusion

Glossary

Abbreviations/Acronyms

Index

CONCLUSION

Paramount among the long-term needs that have to be met by all nuclear operations are reduction of the risks to the environment and to human health, environmental remediation of affected areas, safe disposition of radioactive waste and fissile materials (in particular of weapons grade plutonium), and better public acceptance of necessary nuclear activities.

Needs that are specific to the future development of separation science and technology have been emphasized by the contributors to this book and were also noted by the participants of the workshop on which the book is based and during a subsequent summary meeting of workshop presenters and organizers:

- New separation technologies are needed to minimize wastes in current and future nuclear operations. Environmental safety should be improved, and remediation of areas affected by past operations should continue. Partitioning of wastes is needed to separate the long-lived isotopes, while special waste forms are needed to dispose of these isotopes safely.

- Separations and waste from production are sequential operations; therefore, development of both technologies should proceed with interaction and cooperation between separation experts and waste form experts.

- The issues facing separations developments are primarily economic, not technical. The key to the development of new processes, therefore, is to reduce their cost.

- Transmutation and other future possibilities should be considered as complementary to vitrification and burial. The long-lived isotopes should be separated and stored, not buried, so that transmutation and other options may be used in the future to dispose of these isotopes.

- Fewer courses and research opportunities in nuclear chemistry are being offered at universities worldwide, which could become a serious problem for the future development of better separation methods. A recent successful summer program in the U.S. to introduce students to nuclear chemistry and build their interest in that science could be considered as a model for future programs of this kind.

- Public acceptance is among the most important factors which could affect the future development of nuclear technologies. The public view of nuclear issues will play a strong role in continued R&D on the technical issues. Therefore, an international effort to attract support for nuclear operations would be of great value. Long-term research toward reducing the risk of the separations in nuclear waste operations, not just reducing their cost, is needed to restore and improve public confidence.

- Funding organizations must be made aware that the best available methods are not always considered in solving environmental problems caused by nuclear operations. Disposing of all nuclear wastes in a single final waste form such as borosilicate glass, for example, may make it impossible to make use of promising alternative options in the future.

- The benefits of international cooperation in an area as complex and universal as separation science and technology are self-evident. Many separation technologies are used or are being considered for future use in countries other than the country where they were initially developed. Nuclear waste management is an international problem that cannot be solved by national solutions alone.

The contributors to this volume are aware of the difficulties and the subjective aspects of analyzing future directions in nuclear separation science and technology. Their articles are intended to encourage experts in the field as well as a wider audience to engage in further discussion of these issues.

GLOSSARY

actinides—a group of elements with atomic numbers from 89 to 103, inclusive.

activity (radioactivity)—the number of radioactive transformations that take place in a given period of time; measured in becquerel (Bq) or curie (Ci).

aquifer—a water-bearing formation below the surface of the Earth.

aqueous process—chemical separation process by solvent extraction, e.g., PUREX, TRUEX, and TRAMEX.

back-end of the fuel cycle—the part of the fuel cycle that includes spent fuel storage, fuel reprocessing, mixed-oxide fuel fabrication, and waste management including spent fuel disposal.

becquerel (Bq)—the SI unit of radioactivity defined as one decay per one second.

benchmark—established reference value or standard against which health-based or environmental comparisons may be made.

blanket—material that is put next to a reactor core in order to expose it to the neutron flux.

borosilicate glass—a supercooled liquid based on a random lattice of silica tetrahedra, modified with boron and other cations, used as an immobilization matrix for radioactive waste.

ceramic materials—solid materials of a crystalline structure, usually containing SiO_2 and metal oxides.

curie (Ci)—a unit of radioactivity that is defined as 3.7×10^{10} decays per second, i.e., 1 Ci = 37 MBq; two subsidiary units, representing smaller amounts of active material are the millicurie and the microcurie.

decommissioning—the work required for the planned permanent retirement of a nuclear facility from active service.

decontamination—the removal of radioactive contaminants with the objective of reducing the residual radioactivity level in or on materials, persons, or the environment.

embedding—a process of putting solid or liquid waste into a matrix to form a heterogeneous waste form.

eutectic mixture—a mixture of two solids with the lowest possible freezing point.

front-end of the fuel cycle—mining, milling, enrichment, and fabrication of nuclear fuel.

fuel reprocessing plant—facility where spent-fuel elements are dissolved, waste materials are removed, and reusable materials are segregated.

geological repository—a final disposal facility located deep underground in a stable formation such as salt or granite.

glass ceramic—the product resulting after a glass has been transformed into a crystalline material by a controlled process such as heating.

gray (Gy)—the SI unit of absorbed radiation dose; one gray is equal to the absorbed dose of any ionizing radiation which is accompanied by the liberation of one joule (J) per kilogram (kg) of absorbing material, 1 Gy = 1 J/kg.

half-life ($t_{1/2}$)—the time required for decay of one half of any amount of radioactive isotope.

high-level waste (HLW)—highly radioactive waste resulting from chemical processing of spent nuclear fuel and irradiated target assemblies; usually a combination of TRUs and fission products.

interim storage—storage of radioactive materials such that: (1) isolation, monitoring, environmental protection, and human control are provided; and (2) subsequent action involving treatment, transport, and disposal or reprocessing can be accomplished.

International Science and Technology Center (ISTC)—coordinates the efforts of governments, international organizations, and private sector industries to provide weapons scientists from Commonwealth of Independent States countries with opportunities to redirect their talents to peaceful science.

leaching—(1) extraction of a soluble substance from a solid by a solvent with which the solid is in contact; (2) gradual dissolution/erosion of emplaced solid waste or chemical therefrom, or removal of sorbed material from the surface of a solid or porous bed.

ligand—a chemical group bound to a central metal atom by a coordinate bond.

low-level waste (LLW)—radioactive waste not classified as HLW, TRU, spent nuclear fuel or byproduct material and acceptable for disposal in a licensed land disposal facility; typically includes discarded radioactive materials such as rags, construction rubble, and glass that are only slightly or moderately contaminated.

medium-level waste (MLW)—medium-level waste, or intermediate-level waste.

molten salt reactor- a type of reactor that uses molten salts of uranium for fuel and coolant.

mill tailings— large volumes of material left from uranium mining and milling. While this material is not categorized as nuclear waste, tailings are of concern both because they emit radon and because they are usually contaminated with toxic heavy metals, including lead, vanadium, and molybdenum.

matrix—in waste management, a nonradioactive material used to immobilize radioactive waste in a monolithic structure.

mixed waste—waste that contains both radioactive and chemically hazardous materials. All high-level and transuranic wastes are managed as a mixed waste. Some low-level waste is mixed waste.

monitoring—the methodology and practice of measuring levels of radioactivity either in environmental samples or en route to the environment.

occupational exposure— the exposure received by a worker during a period of work.

partitioning— separation of nuclear waste materials according to isotopic or elemental content.

plume—material spreading from a source into the surrounding medium, e.g., from a smokestack into the air or from a leaking nuclear waste container into the ground.

rad (radiation absorbed dose)— the absorbed dose of any ionizing radiation that is accompanied by the liberation of 100 ergs of energy per gram of absorbing material, i.e.,1 rad = 10^{-2} Gy.

radionuclide migration—the movement of radionuclides through a medium due to fluid flow or by diffusion.

rem (roentgen equivalent man)—the historic unit that measures the biological damage resulting from the radiation exposure of humans (now replaced by the SI unit sievert (Sv), 1 rem = 10^{-2} Sv). One rem is defined as the absorbed dose of any radiation that is equivalent in biological damage due to the absorption of one rad of gamma rays; 1 rem = 1rad × RBE, where RBE is the relative biological effectiveness of nuclear radiation.

reprocessing— processing of spent reactor fuel to recover the unused fissionable material.

roentgen (R)—the historic unit of absorbed radiation dose. One roentgen is equal to the absorbed dose of gamma rays that is accompanied by liberation of 86 ergs of energy per 1 gram of air, as measured at 0° C and standard atmospheric pressure (760 mm Hg); or the quantity of gamma rays which produces one electrostatic unit of either sign in 1 cm^3 of dry air, as measured at 0° C and standard atmospheric pressure (760 mm Hg).

sidereophores—agents produced by bacteria that can bind metal ions such as Fe(II) strongly, using oxygen donors.

sievert (Sv) the SI unit for radiation biological damage and defined as the absorption dose of any radiation that results in the same biological damage as the absorption of one gray (Gy) of gamma radiation dose; 1 Sv=100 rem.

site characterization—process of systematic sample collection and analysis to determine the extent, magnitude, and distribution of site chemicals or other potential hazards.

spent fuel—irradiated fuel units that can no longer sustain a chain reaction.

synroc— a group of specially formulated zirconium-based ceramics originally developed by Australian scientists for immobilizing high-level waste.

transuranium (TRU) waste—waste that contains a emitting TRU elements with a $t_{1/2}$ greater than 20 years and a total activity of more than 100 nanocuries per gram of waste; the great majority results from weapons production processes and contains plutonium.

transmutation of nuclear waste—conversion of long-lived radionuclides into shorter-lived or stable ones by means of nuclear processes in reactors or accelerators.

uranium-mill tailings—large volumes of material left from uranium mining and milling. While this material is not categorized as waste, tailings are of concern both because they emit radon and because they are usually contaminated with toxic heavy metals, including lead, vanadium, and molybdenum.

vitrification—a process of converting materials into glass-like form.

waste form—the physical and chemical form of the waste materials (e.g. , liquid, in concrete, in glass) without any packaging.

ABBREVIATIONS / ACRONYMS

ADS	accelerator-driven system
ANL	Argonne National Laboratory
APT	accelerator production of tritium
ATW	accelerator transmutation of [nuclear] waste
ADTT	accelerator-driven transmutation technologies
An	actinide elements
At. No.	atomic number
Bq, pBq	becquerel, pico-becquerel
BN	fast breeder reactor (USSR)
BNL	Brookhaven National Laboratory
BNOC	bifunctional neutral organophosphorus compounds
CMPO	carbamoyl phosphine oxide
ChCoDic	chlorinated cobalt dicarbollide
CSEX	cesium extraction process
DCC	chlorine cobalt dicarbollide
DIAMEX	diamide extraction process
DOE	U.S. Department of Energy
DTPA	diethylenetriaminepentacetic acid
EA	Energy Amplifier
EU	European Union
FP	fission products
HLW	high-level radioactive waste
HTGR	high-temperature gas-cooled reactor
ISTC	International Science and Technology Center
INEEL	Idaho National Environmental and Engineering Laboratory
JAERI	Japan Atomic Energy Research Institute
KRI	V.G. Khlopin Radium Institute, St. Petersburg, Russia
LAMPRE	Los Alamos Molten Plutonium Reactor Experiment
LANSCE	Los Alamos Neutron Science Center

LAMPF	Los Alamos Meson Physics Facility
LLFP	long-lived fission products
LLW	low-level radioactive waste
Ln	lanthanide elements
LWR	light-water reactor
MIBK	methyl isobutylketone
MINATOM	Ministry of Atomic Energy of the Russian Federation
MOX	mixed oxide fuel—a blend of uranium and plutonium dioxides
MYRRHA	Multipurpose hYbrid Research Reactor for Harmful Actinides
M.A.	minor actinides
NFC	nuclear fuel cycle
NPP	nuclear power plant
PA	Production Association (Russia)
PAN	polyacrylonitrile
PEG	polyethylene glycol
P/T	partitioning/transmutation
PUREX	plutonium and uranium recovery by extraction
REDOX	oxidation-reduction chemistry process
REE	rare-earth elements
SAT	systematic approach to training
SLFP	short-lived fission products
SNS	spallation neutron source
SREX	strontium extraction process
TALSPEAK	trivalent actinide/lanthanide separations by phosphorus extractants from aqueous complexes
TBP	tributylphosphate
TPE	transplutonium elements
TRU	transuranic elements
TRAMEX	transuranium element extraction (separation process of An(III, IV)) from Ln(III) using alkylamines or alkylammonium salts extractants
TRUEX	transuranium element extraction using phosphanates
UNSCEAR	United Nations Scientific Committee on the Effects of Atomic Radiation
UNEX	universal extraction

INDEX

A

accelerator-driven transmutation. *See* transmutation
Accelerator Transmutation of Waste Program, 79
accelerators, 77
    See also transmutation
accidents at nuclear power plants, 34
acetohydroxamic acid, 39–40
actinide extraction
    alkaline media extraction, 12
    cobalt dicarbollide process, 12, 62
    ferrocyanide, 13, 41, 42
    ion exchange, 12, 13
    membranes, 13
    natural agents, 13–14
    organophosphorus compounds, 11, 39–41,
        59–61
    soft donor complexes, 11–12
    trivalent form in separation, 9–10, 11
    universal extractant, 63
    See also specific elements and *groups*
        (e.g., transplutonium extraction)
adsorption processes, 12–13
Aliquat-336, 12
alkylpyrocatechols, 12
americium, 40, 41, 42, 43, 63–64
aqueous process, 4–6, 7–12, 37–40, 59–64
Argonne National Laboratory, 8, 11, 52, 80
atmospheric nuclear tests, 19, 30–33, 34

B

b-diketonate ligands, 11, 51
background radiation, 30, 69
Becquerel, H., 3
bidentate organophosphorus compounds, 11, 40–41
    See also organophosphorus compounds

biosphere
    contaminant sources, 17, 19, 20, 22, 30–37
    dose calculations, 69–70
    monitoring, 41–44
    radioactive waste interactions, 20
Bismuth Phosphate process, 4–5
breeder reactors, 51–52, 52–53
Bryansk Oblast, 34

C

cadmium and pyrochemical processes, 53
calcine, 59
carbamoylmethylphosphine oxide, 8, 9, 39–40, 61,
    63–64
Central Black Soil Regions, 34
ceramic materials, 13
CERN (European Organization for Nuclear
    Research), 77–79
cesium, 31
cesium extraction methods, 11, 12, 26, 37–38, 62,
    63–64
Chelyabinsk. *See* Mayak Industrial Association
Chernobyl Nuclear Power Plant, 33, 34
chitin, 14
chloride, 10, 51
chlorinated cobalt dicarbollide process, 37–38,
    62–64
Choppin, Gregory R., 16, 55
cladding removal, 51
cobalt dicarbollide process, 12, 37–38, 62–64
Cold War. *See* Russia; United States
collective doses, 68, 70
commercial nuclear explosions, 33
commercial radionuclide recovery, 38, 61, 62–63
cooperation, international, 57–65, 86